Animal Husbandry and Livestock Management

Animal Husbandry and Livestock Management

Editor: Ashlie Archer

RCALLISTO REFERENCE

www.callistoreference.com

Callisto Reference,
118-35 Queens Blvd., Suite 400,
Forest Hills, NY 11375, USA

Visit us on the World Wide Web at:
www.callistoreference.com

ISBN: 978-1-64116-102-2 (Hardback)

Cataloging-in-Publication Data

Animal husbandry and livestock management / edited by Ashlie Archer.
    p. cm.
Includes bibliographical references and index.
ISBN 978-1-64116-102-2
1. Animal culture. 2. Domestic animals. 3. Livestock. I. Archer, Ashlie.
SF61 .A55 2019
636--dc23

Table of Contents

Permissions

List of Contributors

Index

Preface

This book has been an outcome of determined endeavour from a group of educationists in the field. The primary objective was to involve a broad spectrum of professionals from diverse cultural background involved in the field for developing new researches. The book not only targets students but also scholars pursuing higher research for further enhancement of the theoretical and practical applications of the subject.

Animal husbandry and livestock management are two interrelated fields that study the rearing and breeding of livestock to obtain farm labor, meat, wool, eggs, etc. Animals are also raised for specific purposes such as producing vaccines and antiserums. These fields have multiple branches such as poultry farming, dairy farming, aquaculture, etc. Animal husbandry is also concerned with varied aspects of animal feeding, their breeding, health and welfare. Some of the popular practices of this domain include intensive animal farming, cross-breeding, artificial insemination, etc. This book attempts to understand the multiple branches that fall under animal husbandry and livestock management. It presents researches and studies performed by experts across the globe. This book will prove immensely beneficial to students and researchers in these fields.

It was an honour to edit such a profound book and also a challenging task to compile and examine all the relevant data for accuracy and originality. I wish to acknowledge the efforts of the contributors for submitting such brilliant and diverse chapters in the field and for endlessly working for the completion of the book. Last, but not the least; I thank my family for being a constant source of support in all my research endeavours.

Editor

PREVALENCE OF THEILERIOSIS AND BABESIOSIS IN CATTLE IN SIRAJGANJ DISTRICT OF BANGLADESH

Md. Abdullah-Al-Mahmud[1*], SM Shariful Hoque Belal[2] and Md. Alamgir Hossain[3]

[1]Additional Veterinary Surgeon, Upazila Livestock Office, Ullapara, Sirajganj; [2]Veterinary Surgeon, District Veterinary Hospital, Sirajganj; [3]Department of Medicine, Sylhet Agricultural University, Sylhet, Bangladesh

*Corresponding author: Md. Abdullah-Al-Mahmud; E-mail: dr.mahmud04@gmail.com

ARTICLE INFO

ABSTRACT

Key words

Prevalence
Theileriosis
Babesiosis
Cattle

A study was conducted to investigate the prevalence of theileriosis and babesiosis in cattle in Sirajganj district of Bangladesh during the period of December 2013 to November 2014. During one year study period a total of 395 cattle were examined, 23 and 8 were found to be infected with *Theileria* spp. and *Babesia* spp., respectively. On Geimsa stained blood smear examination, it was observed that the overall prevalence of theileriosis and babesiosis in cattle were recorded as 5.82% and 2.27% respectively. The effect of age, sex, breed and season was observed in cattle during this study. The highest prevalence of theileriosis (7.25%) and babesiosis (3.10%) was reported in the older cattle (>3 years of age) and the higher prevalence was observed in female (6.66% and 2.59%, respectively) than male (4.0% and 1.60% respectively). All crossbred cattle was showed higher prevalence than local cattle. The prevalence of theileriosis was noticed as the highest in the rainy season (6.25%) in relation to summer (5.83%) and winter (5.05%) season. But the prevalence of babesiosis was ranked the highest in summer season (2.50%) followed in rainy (2.27%) and winter (2.02%) season that was insignificant. From the study it was evident that cattle were infected with the organisms and caused a heavy economic loss which will assist to take necessary preventive measurements.

INTRODUCTION

Babesiosis and Theileriosis are febrile, tick-borne haemoprotozoan diseases of cattle, caused by *Babesia* spp. and *Theileria* spp. respectively, wide spread in tropical and sub-tropical regions of the world. Among tick-borne hemoprotozoan parasites of vertebrates *Theileria* and *Babesia* are the species that have a major effect on livestock health (Mehlhorn and Schein, 1984). Babesiosis, like most haemoparasites, has generally been shown to cause destruction of red blood cells resulting in anaemia, jaundice, anorexia, weight loss and infertility (Ellis et al., 2003); and *Theileria* sp.is a common protozoan parasite of cattle transmitted by *Hyalomma* sp. ticks, responsible for tropical theileriosis, a disease which has been reported from various parts of the world (Oliveira et al., 1995; Durrani et al., 2010; Tavassoli et al., 2011). Tick-borne diseases cause substantial losses to the livestock industry throughout the world (Ananda et al., 2009; Kakarsulemankhel, 2011) as these have got a serious economic impact due to obvious reason of death, decreased productivity, lowered working efficiency (Uilenberg, 1995), increased cost for control measures (Makala et al., 2003) and limited introduction of genetically improved cattle in an area (Radostits et al., 2000).

Bangladesh is usually hot and humid except in winter, and the climatic condition of Bangladesh is very conducive to a wide variety of parasites as well as ticks (Razzak and Shaikh, 1969) which have been recognized as the notorious threat due to severe irritation, allergy and toxicosis (Niyonzema and Kiltz, 1986). Prevalence of blood protozoa such as *Babesia bigemina, Theileria annulata, Theileria* mutans has been reported in animals of Bangladesh (Ahmed, 1976; Samad and Gautam, 1984); where Alim et al. (2012) observed that the prevalence of theileriosis and babesiosis was 4.62% and 1.85%, respectively in Noakhali district, and 8.33% and 2.78%, respectively in Khagrachori district. Chowdhury et al. (2006) recorded the prevalence of *Babesia bigemina* infection in cattle was 3.30% in Sirajganj sadar area of Bangladesh. Outbreaks of clinical illness do not occur uniformly in tick-infected areas due to the differences in genetic or age resistance of the cattle, variations in tick populations, numbers of infected ticks and established control methods (de Castro, 1997). Kamani et al. (2010) observed that the higher prevalence was found in adult than young cattle. Sarkar (2007) and Khattak et al. (2012) also found that the higher prevalence was showed in female than male. Prevalence of the disease was higher in croosbreed cattle than local cattle which was reported by Siddiki et al. (2010) and Esin et al. (2012). Seasonal variation on tick infestation was reported by Zahid et al., (2005) and Sanjay et al. (2007) where they reported that the prevalence of tick infestation was significantly more during the rainy season. Sirajganj is the most important dairy belt of Bangladesh and the density of cattle population of this district is high. The climatic condition and geographical location of the areas might favor the growth and multiplication of of ticks which act as natural vectors of babesiosis and theileriosis that causes one of the major veterinary problems affecting livestock industries. Considering the above situations, the present research was undertaken to know the prevalence of babesiosis and theileriosis of cattle in Sirajganj district of Bangladesh.

MATERIALS & METHODS

Geographical location of study area

Sirajganj district is situated in Rajshahi Division of Bangladesh; its geographical coordinates are 24° 27' 0" North, 89° 43' 0" East. Sirajganj has an area of 2,498 sq.km (964 sq miles) including reverine areas, and it represents around 1.7 percent of the total area of Bangladesh. It ranks 3rd in size among the eight districts of Rajshahi division and 25th among the 64 districts of Bangladesh. It contains 9 upazilas namely SirajganjSadar, Shahjadpur, Ullapara, Kamarkhand, Belkuchi, Chouhali, Raigonj, Tarash and Kazipur.The annual average temperature of the district reaches a maximum of 34.6 °C, and a minimum of 11.9 °C. The annual rainfall is 1610 mm (63.4 in).

Experimental animals and duration of study

This research work was conducted at the Field Diseases Investigation Laboratory (FDIL), Sirajganj on the clinical cases in cattle during the period from December 2013 to November 2014. During one year study period, a total of 395 case of sick and suspected cattle were studied that was informed by farmers, Veterinary Field Assistant, Veterinary Surgeon and Upazila Livestock Officer of respective upazila of Sirajganj district.

Only 23 and 9 cattle were infected by *Theileria* spp and *Babesia* spp. among 395 suspected cases. Date, age, sex, breed and complaint of the owner of all studied animal were noted in the registered book. All this information and data were collected from the disease register book of the FDIL, Sirajganj.

Methods followed for diagnosis

The history and physical examination of each of the patient were carried out for the cattle are briefly described below:

History/Anamnesis

History of patient like (a) Date of examination, (b) Signalment (client and patient) identification, (c) Chief complaint, (d) Patient illness, (e) Past medical history were included. In addition, the complete medical history like (a) Family medical history, (b) Vaccination history, (c) Travel history, (d) Diet history, (f) Environmental history, (g) Birth history (h) Potential source of intoxication were investigated.

Physical examination

Physical examination was done by visual inspection, pulse & respiration rate and rectal temperature. Examination of the different organs and systems of the body was carried out by using the clinical methods of palpation, percussion and auscultation.

Sample collection and examination

3-5 ml blood samples were collected from the jugular vein of the clinically suspected animals in EDTA containing vacutainers and transported to FDIL, Sirajganj in ice bags for microscopic examination following the method of Adam, Paul and Zaman. Briefly, a thin blood smear was prepared from each blood sample, air dried and fixed in methanol for 2-3 minutes. Staining was done in 5% Giemsa's stain and rinsing was performed in two changes of distilled water buffered to pH 7.2, then examined under microscope (100x) with immersion oil for the identification of blood parasites as described by Soulsby (1982).

Statistical analysis

Data were entered in Microsoft Excel 2007 and transferred to R 2.14.2 (The R Foundation for Statistical Computing, Vienna, Austria). Descriptive statistics were obtained using Data Mining package of the software R 2.14.2

RESULTS AND DISCUSSION

Overall prevalence of theileriosis and babesiosis in cattle

Overall prevalence of theileriosis and babesiosis of cattle in Sirajganj district is shown in table 1. The overall prevalence of Theileriosis and Babesiosis of cattle in Sirajganj district was 5.82% and 2.27% respectively. The present study supports with the report of Khattak et al. (2012) who observed that the overall prevalence of Theileriosis on Giemsa stain was 5.20% in Khyber PukhtoonKhwa province (Pakistan). Similar results were also observed by Mohammed Safieldin et al. (2011) who found that the prevalence of *Theileria* species infection in dairy cattle in Omdurman locality, Sudan was found to be 7% for dry cool season while the prevalence of *Babesia* species infection was1%. In Bangladesh, similar findings were also reported by Alim et al. (2012) where he observed that the prevalence of theileriosis and babesiosis was 4.62% and 1.85% respectively in Noakhali district, and 8.33% and 2.78% respectively in Khagrachori district.Babesia infection (2.27%) recorded in this study is similar to the report of Shahidullah (1983) who recorded the prevalence of babesiosis was 2.29% on microscopic peripheral blood smear examination. Samad et al. (1989) and Chowdhury et al. (2006) recorded a comparatively higher (3.28% and 3.30% respectively) prevalence of *Babesia bigemina* infection in cattle of the selected Milk vita project areas of Bangladesh and Sirajganj sadar area of Bangladesh respectively. But Banerjee et al. (1983) detected very high (14.53%) prevalence of *Babesia bigemina* in cattle of Bangladesh.

Table 1. Overall prevalence of theileriosis and babesiosis of cattle in Sirajganj district

Name of Diseases	No. of cattle tested (for both diseases)	No. of positive case	Prevalence (%)
Theileriosis	395	23	5.82
Babesiosis		9	2.27

Age-wise prevalence

Age-wise prevalence of Theileriosis and Babesiosis in cattle is shown in table 2. Age also influences the occurrence of the infections. The highest prevalence of Theileriosis and Babesiosis was found in the age of above 3 years (7.25% and 3.10% respectively), followed in >2-3 years (5.12% and 1.70% respectively) and 6 months – 2 years of age (3.52% and 1.17% respectively).Similar results were found byKamani et al. (2010) who observed higher prevalence in adult than young cattle. The results of present study agree with Islam et al. (2009) who found that prevalence of tick infestation was higher in old cattle than young. Observation of this study also supported by the findings of of Ananda et al. (2009) who reported higher prevalence in animals aged more than 3 years followed by the lower prevalence in 1-2 years of age. Findings of babesiosis in this investigation were supported by the observation of Urquhart et al. (1996) and Annetta et al. (2005) who reported an inverse age resistance of the disease where adult showed more susceptibility than calves. This might be due to rapid immune responses to primary infection by the calves through a complex immune mechanism (Annetta et al., 2005). Endemic instability of the study areas might responsible for frequent infections in adult cattle where newborn calves were protected by colostral immunity (Cynthia et al., 2011). Age resistance, perhaps in combination in some cases with maternal antibodies, is reflected in the reduced number of clinical outbreaks in young animals.

Table 2. Age-wise prevalence of theileriosis and babesiosis in cattle

Age group	No. of cattle tested	Theileriosis		Babesiosis	
		No.	%	No.	%
6 months-2 years	85	3	3.52	1	1.17
>2-3 years	117	6	5.12	2	1.70
Above 3 years	193	14	7.25	6	3.10
Total	**395**	**23**	**5.82**	**9**	**2.27**

Sex-wise prevalence

Sex-wise prevalence of Theileriosis and Babesiosis in cattle is shown in table 3. In case of both Theileriosis and Babesiosis, the higher prevalence was observed in female (6.66% and 2.59% respectively) than male (4.0% and 1.60% respectively). Similar findings were found by Singh et al. (2012) and Khattak et al. (2012) where they separately observed that the prevalence of theileriosis was higher in female than male. This result also agree with the report of Sarkar (2007) who reported the prevalence of ectoparasites (ticks) were significantly higher in female than male. The prevalence of hemoprotozoan diseases in female cattle of this investigation showed uniformity with the report of Kamani et al. (2010). Higher prevalence in female cattle possibly due the fact that they were kept longer for breeding and milk production purpose, supplied insufficient feed against their high demand (Kamani et al., 2010) or variation in sample size.

Table 3. Sex-wise prevalence of Theileriosis and Babesiosis in cattle

Sex group	No. of cattle tested	Theileriosis		Babesiosis	
		No.	(%)	No.	(%)
Male	125	5	4.0	2	1.60
Female	270	18	6.66	7	2.59
Total	**395**	**23**	**5.82**	**9**	**2.27**

Breed-wise prevalence

Breed-wise prevalence of Theileriosis and Babesiosis in cattle is shown in table 4. The highest prevalence of Theileriosis was recorded in HF×L (6.66%) followed by SL×L (6.18%), S× L (5.55%) and local (4.54%). In case of Babesiosis, the highest prevalence was reported in HF×L (2.50%) followed by local (2.27%), S× L (2.22%) and SL×L (2.06%). In the present study, higher prevalence of *Theileria* infections in crossbreed cattle as compared to local cattle was found. Breed differences are also important in the susceptibility of cattle to tick borne diseases. Cattle of European origin like Holstein are usually highly susceptible (Esin et al. 2012). Breed susceptibility of Babesiosis also recorded in this study support the report of Chakraborti A, (2002). Variation in geo-climatic condition, breed, and exposure of vectors and age of the animals might contribute to variable prevalence of hemoprotozoan diseases in the study areas (Muhanguzi et al., 2010). Constant exposure of infections and development of immunity against such infections might responsible for lower prevalence in indigenous cattle (Siddiki et al., 2010). On the contrary, more attention in the management of HF crossbred cattle gives less chance of pre exposure of vectors and develop no or less immunity, resulting frequent occurrence of such diseases (Chowdhury et al., 2006; Ananda et al., 2009; Siddiki et al., 2010).

Table 4. Breed-wise prevalence of Theileriosis and Babesiosis in cattle

Breed	No. of cattle tested	Theileriosis		Babesiosis	
		No.	%	No.	%
Local	88	4	4.54	2	2.27
HF×L	120	8	6.66	3	2.50
SL×L	97	6	6.18	2	2.06
S× L	90	5	5.55	2	2.22
Total	**395**	**23**	**5.82**	**9**	**2.27**

Local=Indigenous cattle, HF = Holstein-Friesian, SL = Sahiwal, S = Sindhi, L= Local cattle

Table 5. Season-wise prevalence Theileriosis and Babesiosis in cattle

Season	No. of cattle tested	Theileriosis		Babesiosis	
		No.	%	No.	%
Summer	120	7	5.83	3	2.50
Rainy	176	11	6.25	4	2.27
Winter	99	5	5.05	2	2.02
Total	**395**	**23**	**5.82**	**9**	**2.27**

Summer: March- June; Rainy: July-October and Winter: November-February

Season-wise prevalence

Season-wise prevalence of Theileriosis and Babesiosis in cattle in the study area is shown in table 5. The prevalence of Theileriosis was ranked the highest in rainy season (6.25%) in relation to summer (5.83%) and winter (5.05%) season. But the prevalence of Babesiosis was observed the highest in summer season (2.50%), followed by rainy (2.27%) and winter (2.02%) season. Occurrence of theileriosis was found in line with the reports of Muhammad et al. (1999) and Zahid et al. (2005). Observation of rainy season of this research was in accordance with the report of Ananda et al. (2009). Radostits et al. (2000) observed that higher incidence of haemoprotozoan diseases were found soon after peck of tick population depending on temperature, humidity, rainfall etc. which might be accounted for higher prevalence of such infections in rainy season of the study. Lower temperature and humidity of winter months were less favorable for the growth and multiplication of tick vectors which might contribute to lower frequency of such diseases in the study population (Muhammad et al., 1999; Zahid et al., 2005). Salih et al. (2008) found the highest number of ticks occur during the rainy season. Sanjay et al. (2007) also reported the seasonal prevalence of tick infestation significantly more during the rainy (24.33%) and summer seasons (21.58%) as compared to the winter season

(4.03%). This variation may be due to geographical location, climatic condition, reproduction of tick and tick population. The present study recorded a higher prevalence of Babesiosis in summer season which was agreed with the report of Rony et al. (2010) where he noticed that the prevalence of ectoparasites/ tick infestation was higher in summer season. The rise of infestation in summer may be due to rise of temperature in late winter leading to gradual increase in the load as well as percentage of infestation in May and June (Roy et al. 2001).

Tick transmitted haemoprotozoan diseases especially babesiosis, theileriosis and anaplasmosis are considered some of the major impediments in the health and productive performance of cattle (Rajput et al., 2005); affecting livestock industries in many parts of the world (Hourrigan, 1979). In some cases, ticks have been reported to cause lowered productivity, mortality (Niyonzema and Kiltz, 1986) and transmit the diseases (Norval *et. al.*, 1984). Haemoprotozoan diseases of cattle have also been recorded from some districts of Bangladesh (Samad et al. (1983); which are one of the major constraints for profitable dairy industry in tropical and subtropical countries including Bangladesh.*Theileria* and *Babesia* species are among the major piroplasms of cattle and small ruminants (Uilenberg, 1995; Gubbels et al., 1999; Oura et al., 2004; Oura et al., 2005); which cause significant economic losses in tropical and sub-tropical regions of the world (Uilenberg, 1995; Kursat et al., 2004; Jongejan and Uilenberg, 2004).

From the above discussion it is concludate that the prevalence of theileriosis and babesiosis is the most common and frequent in Sirajganj district of Bangladesh which may be due to its agro-ecological and geoclimatic conditions, high density of cattle population, lowland/flood plain based area and high density of tick population. So, for the control and eradication of the diseases in cattle population of Bangladesh specially Sirajganj district, epidemiological surveillance is the important aspect along with more attention should be paid towards separation of healthy and infected animals/herds and strictly followed with biosecurity in cattle farm. Reduction of tick population and tick's reproduction is the main issue for control of the diseases. This should be combined with more government intervention in with awareness program should be undertaken involving stakeholders in the livestock industry as well as consumers to avert public health, veterinary public health and economic losses associated with the mentioned diseases.

REFERENCE

1. Ahmed AKNU, 1976. Blood parasites of domestic animals in Bangladesh. Bangladesh Veterinary Journal, 10: 69-71.
2. Ananda KJ, PE Souza and GC Puttalakshmamma, 2009. Prevalence of haemoprotozoan diseases in crossbred cattle in Banglore north. Veterinary World, 2: 15-16.
3. Annetta ZS, Jeremy, Gray, E Helen, Skerrett and G Mulcahy, 2005. Possible mechanisms underlying age-related resistance to Bovine babesiosis. Parasite Immunology, 27: 115-120.
4. Alim MA, D Shubhagata, R Krisna, M Masuduzzaman, S Suchandan, H Mahmudul, AZ Siddiki and
5. MA Hossain, 2011. Prevalence of hemoprotozoan diseases in cattle population of Chittagong Division, Bangladesh. Pakistan Veterinary Journal, 32: 221-224.
6. Banerjee DP, KD Prasad and MA Samad, 1983. Seroprevalence of Babesia bigemina infection in cattle of India and Bangladesh. Indian Journal of Animal Sciences, 53: 431-433.
7. Chakraborti A, 2002. A Textbook of Preventive Veterinary Medicine. 3[rd]end.,Kalyani Publishers, New Delhi. pp. 683.
8. Chowdhury S, MA Hossain, SR Barua and S Islam, 2006. Occurrence of common blood parasites of cattle in Sirajganj sadar area of Bangladesh. Bangladesh Journal of Veterinary Medicine, 4: 143-145.
9. Cynthia M, MA Kahn, S Line, E Susan and BS Aiello, 2011. Marck Veterinary Manual, Online Ed. Merck Sharp & Dohme Corp, a subsidiary of Merck & Co., Inc. Whitehouse Station, NJ, USA.
10. De Castro JJ, 1997. Sustainable tick and tick-borne disease control in livestock improvement in developing countries. Veterinary Parasitology, 77: 77–97.
11. Durrani AZ, N Mehmood and AR Shakoori, 2010. Comparison of three diagnostic methods for Theileria annulatedin Sahiwal and Friesian cattle in Pakistan. Pakistan Journal of Zoology, 42: 467-472.
12. Ellis JT, DA Morrison, MP Reichel, 2003. Genomics and its impact on parasitology and the potential for development of new parasite control methods. DNA. Cellular Biology, 32: 395-403.

13. Esin G, C Ayşe, O Omer, A Aytaç and K Asiye, 2012. Prevalence of Babesia bigemina and Theileria annulata in cattle in Eregli, Turkey. Harran Üniversitesi Veteriner Fakültesi, 1: 18-21

14. Gubbels JM, MAP Devos, J Van Der Weide, LM Viseras, E Schouls, Devries and F Jongejan, 1999. Simultaneous detection of bovine Theileria and Babesia species by reverse line blot hybridization. Journal of Clinical Microbiology, 1782-1789.

15. Hourrigan JL, 1979. Spread and detection of Psoroptic scabies of cattle in the United States. Journal of American Veterinary Association, 175: 1278-1280.

16. Islam MS, SA Rahman, P Sarker, Anisuzzaman and MMH Mondal, 2009. Prevalence and population density of ectoparasitic infestation in cattle in Sirajgonj district, Bangladesh. Bangladesh Research Publications Journal. 2: 332-339.

17. Jongejan F, and G Uilenberg, 2004. The global importance of ticks. Parasitology 129: 3-14.

18. Kakarsulemankhel JK, 2011. Re-description of existing and description of new record of tick [Hyalomma (Euhyalomma) schulzei] from Pakistan. International Journal of Agriculture and Biology, 13: 689–694.

19. Kamani J, A Sannusi, OK Eqwu, GI Dogo, TJ Tanko, S Kemza, AE Takarki and DS Gbise, 2010. Prevalence and significance of haemoparasitic infections of cattle in North-Central, Nigeria. Veterinary World, 3: 445-448.

20. Khattak RM, M Rabib, Z Khan, M Ishaq, H Hameed, A Taqddus, M Faryal, S Durranis, Gillani Qua, R Allahyar, RS Shaikh, MA Khan, M Ali and F Iqbal, 2012. A Comparison of two different techniques for the detection of blood parasite, Theileria annulata in cattle from two districts in Khyber Pukhtoon Khwa province (Pakistan). Parasite, 19: 91-95

21. Kursat A, D Nazir, J Patricia, B Holman and A Munir, 2004. Detection of Theileria ovis in naturally infected sheep by nested PCR. Veterinary Parasitology, 127: 99-104.

22. Mehlhorn H, E Schein, 1984. The piroplasms: life cycle and sexual stages. AdvParasitol. 23: 37–103.

23. Mohammed Safieldin A, AGE Atif, and EH Khitma, 2011. Factors affecting seasonal prevalence of blood parasites in dairy cattle in Omdurman locality, Sudan. Journal of Cell and Animal Biology, 5: 17-19.

24. Muhanguzi D, K Ikwap, K Picozzi and C Waiswa, 2010. Molecular characterization of anaplasma and ehrlichia species in different cattle breeds and age groups in Mbarara district (Western Uganda).International Journal of Animal and Veterinary Advance, 2: 76-88.

25. Makala LH, P Mangani, K Fujisaki and H Nagasawa, 2003. The current status of major tick borne diseases in Zambia. Veterinary Research, 34: 27-45.

26. Muhammad GM, M Saqib, MZ Athar and MN Khan, 1999. Clinico epidemiological and therapeutic aspects of bovine theileriosis. Pakistan Veterinary Journal, 19: 64-69.

27. Niyonzema A and HH Kiltz, 1986. Control of ticks and tick-borne diseases in Burundi. Australian Center for International Agricultural Research, 17:16-17.

28. Norval RAI, BH Fivaz, JA Lawrence and AF Brown, 1984. Epidemiology of tick-borne diseases of cattle in Zimbabwe. Tropical Animal Health and Production, 16: 63-70.

29. Oliveira DC, M Van Der Weide, MA Habela, P Jacquiet and F Jongejan, 1995. Detection of Theileria annulatain blood samples of carrier cattle by PCR. Journal of Clinical Microbiology, 33: 2665-2669.

30. Oura CAL, R Bishop, EM Wampande, GW Lubega and A Tait, 2004. The persistence of component Theileria parva stocks in cattle immunized with the 'Muguga cocktail' live vaccine against East Coast fever in Uganda. Parasitology, 129: 27-42.

31. Oura CAL, BB Asiimwe, W Weir, GW Lubega and A Tait, 2005. Population genetic analysis and Substructuring of Theileria parva in Uganda. Mol. Biochem. Parasitol. 140: 229-239.

32. Razzak A and H Shaikh, 1969. A survey on the prevalence of ticks on cattle in East Pakistan. Pakistan Journal of Veterinary Science, 3: 54-60.

33. Rajput ZI, Hu Song-hua, AG Arijo, H Habib and K Khalid, 2005. Comparative study of Anaplasma parasites in tick carrying buffaloes and cattle. Journal of Zhejiang University Science, 6: 1057-1062.

34. Radostits OM, DC Blood and CC Gay, 2000. Veterinary Medicine: A text book of disease of cattle, sheep, pigs, goats and horse. 9th Ed, BaillereTindall Publication, London, pp: 1172-1173, 1289-1290.

35. Rony SA, MMH Mondal, N Begum, MA Islam and S Affroze, 2010. Epidemiology of ectoparasitic infestations in cattle at Bhawal forest area, Gazipur. Bangladesh Journal of Veterinary Medicine, 8: 27–33.

36. Roy AK, MH Rahman, S Majumder and AS Sarker, 2001. Ecology of ticks and tick-borne blood protozoa in Madhupur Forest Area, Tangail. Bangladesh Veterinary Journal, 17: 90-97.

37. Samad MA and OP Goutam, 1984. Prevalence of Theileria annulata infection in cattle of Bangladesh. Indian Journal of Parasitology, 7: 61-63.

38. Samad MA, SA Bashar, M Shahidullah and MU Ahmed, 1989. Prevalence of haemoprotozoan parasites in cattle of Bangladesh. Indian Journal of Veterinary Medicine, 13: 50-51.

39. Samad MA, 2000. An overview of livestock research reports published during the twentieth century in Bangladesh. Bangladesh Veterinary Journal, 34: 53 -149.

40. Sanjay K, KD Prasad and AR Deb, 2007. Seasonal prevalence of different ectoparasites infecting cattle and buffaloes. Journal of Research, 16: 159-163.

41. Sarkar M, 2007. Epidemiology and pathology of ectoparasitic infestation in Black Bengal Goats in Bangladesh. M.Sc. thesis. Department of Parasitology, Bangladesh Agricultural University, Mymensingh.

42. Salih DA, II Julia, SM Hassan, AM El-Hussain and F Jongejan, 2008. Preliminary Survey of ticks (Acari: Ixodidae) on Cattle in Central Equatoria State, Southern Sudan.Onderstepoort Journal of Veterinary Research, 75: 47-53.

43. Shahidullah, 1983. Studies on haemoprotozoan disease of goats and cattle with their vector ticks. M. Sc. Thesis, Submitted to the Department of Medicine, Bangladesh Agricultural University, Mymensingh.

44. Singh NK, S Harkirat, Jyoti, M Haque, SS Rat, 2012. Prevalence of parasitic infections in cattle of Ludhiana district, Punjab. Journal of Parasitic Diseases, 36: 256-259.

45. Siddiki AZ, MB Uddin, Hasan, MF Hossain, MM Rahman, BC Das, MS Sarker and MA Hossain, 2010. Coproscopic and haematological approaches to determine the prevalence of Helminthiasis and protozoan diseases of Red Chittagong Cattle (RCC) breed in Bangladesh. Pakistan Veterinary Journal, 30: 1-6.

46. Tavassoli1 M, M Tabatabaei, BE Nejad, MH Tabatabaei, A Najafabadi and SH Pourseyed, 2011. Detection of Theileria annulata by the PCR-RFLP in ticks (Acari, Ixodidae) collected from cattle in west and North West Iran. Acta Parasitologica, 56: 8-13.

47. Uilenberg G, 1995. International collaborative research: significance of tick-borne hemoparasitic diseases to world animal health. Veterinary Parasitology, 57: 19-41.

48. Zahid IA, M Latif and KB Baloch, 2005. Incidence and treatment of theileriasis and babesiasis. Pakistan Veterinary Journal, 25: 137-139.

2

COMPARATIVE EFFICACY OF ALCOHOLIC EXTRACTS OF BLACK PEPPERS (*Piper nigrum*) AND CHUTRA LEAVES (*Urtica dioica*) WITH ESB3 AGAINST COCCIDIOSIS IN CHICKENS

Bayzer Rahman, Tahmina Begum, Yousuf Ali Sarker, Md. Quamrul Hassan Sukumar Saha[1], Mahmudul Hasan Sikder and Md. Abdul Awal*

Department of Pharmacology, Faculty of Veterinary Science, Bangladesh Agricultural University, Mymensingh-2202, Bangladesh
[1]Department of Microbiology and Hygiene, Faculty of Veterinary Science, Bangladesh Agricultural University, Mymensingh-2202, Bangladesh

*Corresponding author: Md. Abdul Awal, E-mail: dmawalbau@gmail.com

---

ARTICLE INFO

Key words

Efficacy
Medicinal plants
Coccidiosis
Chicken

ABSTRACT

The present study is undertaken to compare the efficacy of alcoholic extracts of Black Peppers (*Piper nigrum*) and Chutra leaves (*Urtica dioica*) with a patent drug Esb3 against coccidiosis in chicken. 16 Fayoumi breed chickens were collected from a local farm and divided into four groups; A, B, C and D, each consisting of four chickens, Group A (control), Group B (alcoholic extracts of Black Peppers @ 9ml/kg bd wt.), Group C (alcoholic extracts of Chutra leaves @ 9ml/kg bd wt.) and Group D (Esb3 @ 1 ml/ liter drinking water). All the treated chickens were kept under close observation for 18 days and data was collected at 3 days interval. In group B, two chickens died within 4-7 days of treatment and in group C one chicken died on 5th day of medication. All the chickens of control group died within 5-7 days of medication. Oocyst was counted for per gram of feces in all groups. Biochemical parameters like SGPT and haematological parameters like Total erythrocyte count (TEC), Haemoglobin count (Hb), Packed cell volume (PCV), Erythrocyte sedimentation rate (ESR) were determined. There were significant decrease in oocyst count in group B and C in compared to control and very few oocysts were present in faeces of 6 days onward of medication. All the chickens were survived in group D and oocyst started to disappear in the faeces from 3rd day onward of medication. Our study suggests Both Black Peppers and Chutra are effective against coccidiosis in chicken and Chutra is more effective than Black peppers.

INTRUDUCTION

Coccidiosis is the most important protozoan disease affecting the poultry industry worldwide. Control of poultry coccidiosis is presently based on managerial skills and the use of prophylactic coccidiostatic drugs (Tewari and Maharana, 2011). It is a ubiquitous intestinal protozoan infection of poultry seriously impairing the growth and feed utilization of infected animals. Conventional disease control strategies rely heavily on chemoprophylaxis, which is a tremendous cost to the industry. The economic significance of coccidiosis is attributed to decreased production (higher feed conversion, growth depression and increased mortality) and the costs involved in treatment and prevention. Worldwide, the annual costs inflicted by coccidiosis to commercial poultry have been estimated at 2 billion euro, stressing the urgent need for more efficient strategies to control this parasite (Peek and Landman, 2011). Numerous anticoccidial drugs have been introduced since 1948, when sulphaquinoxaline and nitrofurazone were first approved by the American Food and Drug Administration (Conway DP, 2007). Existing vaccines consist of live virulent or attenuated *Eimeria* strains with limited scope of protection against an ever-evolving and widespread pathogen. The continual emergence of drug-resistant strains of *Eimeria*, coupled with the increasing regulations and bans on the use of anticoccidial drugs in commercial poultry production, urges the need for novel approaches and alternative control strategies (Dalloul and Lillehoj, 2005).

The ecological factors prevailing in Bangladesh are highly conducive for the survival, multiplication and perpetuation of poultry parasites of which coccidiosis is encountered in chicken. Mortality in young birds which varies from 25-90% is predominant factor of economic loss. Most of the farmers are very poor and cannot afford to buy modern drugs for the treatment various diseases due to unavailability and higher price of the drugs. If the farmers can use the traditional medicinal plants for the treatment of various animal and poultry diseases it will be very much helpful for them and for the overall improvement of livestock. In Bangladesh very few works have been done to explore the possibilities of utilizing the indigenous plants in poultry coccidiosis. Black peppers have been shown as a growth promoter in poultry (Abou-Elkhair, Ahmed, and Selim, 2014) and sulfachlorpyrisine-Na (Esb3) was active against coccidiosis (Penev and Lozanov, 1983). However, resistance has been reported for Esb3 in poultry (Harfoush et al., 2010). No data yet available about the efficacy of Black Peppers and Chutra leaves in coccidiosis. Therefore, the aim of this study was to evaluate the efficacy of plant extracts (Black Peppers and Churta leaves) compared to ESB3 drug in poultry coccidiosis.

MATERIALS AND METHODS

The experiment was conducted in the Department of Pharmacology and Department of Physiology, Faculty of Veterinary Science, Bangladesh Agricultural University, Mymensingh.

Experimental Chickens

Total 16 chickens of Fayoumi breed were collected from a local farm. The chicks were 3-4 weeks of age and weighing 250-300 gm. The experimental poultry shed was properly brushed with broom and then washed by forced water using a hosepipe. The room was disinfected with bleaching powder and it was left for 7 days. After the interval, the shed was again disinfected with Virkons (Antes International Limited, England). Additionally the poultry shed was also fumigated by formalin and potassium permanganate for a period of 24 hours for disinfection.

Collection of Coccidia Affected Birds and Isolation of Oocyst and Infection

Four chickens suffering from coccidiosis were collected from a private farm. The ceca of coccidian infected chickens were separated and opened with scissors and forceps and the fecal contents were taken out in a petri dish containing water. The content was then stirred for 20 minutes and filtered. 50 ml of filtrate was taken in a glass jar. A drop of filtrate was examined under the microscope to observe oocyst. In 50 ml of filtrate 1 gm of potassium dichromate was added and mixed thoroughly and was kept for 24 hours for sporulation. After 24 hours, a drop of filtrate was examined under the microscope to observe sporulatedoocyst. The oocysts were counted by McMaster Egg counting method.

All experimental birds were divided into four groups and were kept in four different cages. Oocyst containing filtrate was fed to all birds of the groups at the dose rate of 20 ml / bird. All the birds become infected within 3-5 days. The infection of the bird was confirmed by observing the visible symptoms and also by faecal egg count per gm by McMaster Egg counting method.

Alcoholic Extracts of Black Peppers and Chutra

The leaves of Chutra and whole of black Peppers were collected in fresh condition. They were washed thoroughly with fresh water. These were cut into small pieces, sundried and grinded in a grinding machine. The powdered sample was measured with a balance. 100 gm of each sample was taken and preserved in air tight bottle separately. Alcoholic extracts of Black Peppers and Chutra were obtained by Soxhlet method (250 ml of alcohol was used for 100 gm of powdered sample).

Determination of Biochemical and Hematological Parameters

Biochemical parameters like serum glucose and SGPT (serum glutamic pyruvic transaminase) were determined by using autoanalyser (Model No. Reflotron M-06). Haematological parameters like Total Erythrocyte Count (TEC), Haemoglobin count (Hb), Packed Cell Volume (PCV), Erythrocyte Sedimentation Rate (ESR) was determined. The parameters were determined as per method cited by (Coffin, 1955).

Statistical Analysis

The data was analysed statistically between normal and treated values by student's t test and a p value of ≤ 0.5 was considered significant.

RESULT AND DISCUSSION

Effect of Black Peppers on Oocyst Count

The data obtained from oocyst count of per gm of feces showed Black Peppers reduced the oocyst count significantly as compared to control. Data are presented in Table 1. Data was collected up to 12 days during the treatment period and up to 6 days post treatment period at 3 days interval. All the chickens of control group died within 4-8 days of medication. Two chicks of group B died within 4-7 days of medication. Rest two chicks became apparently normal within 7-10 days of treatment. At day 0, oocyst count of per gm of feces in control was 5272.50±31.98 compared to 5212.40 ±94.20 in Black peppers treated group (group B). Oocyst count started to fall 3 days after treatment in group B. However, in control, oocyst count never reduced. All the chickens in group A (control) died after 9 days of treatment. Therefore, the data obtained could be compared with the control up to 9 days during treatment period. On day 9, average oocyst count in group B was 580 compared to 6127 in control group. This observation is statistically significant. While the efficacy of Black Peppers compared with standard Esb3 treated group (group D), the pattern of reducing oocyst count is identical in both group B and D. In both groups, oocyst count started to fall 3 days after treatments. At the end of treatment, oocyst count in group B was 130 compared to 52±8.62 in group D. Observations in post treatment period at day 6 showed oocyst count in group B was 37 while in group D was 14±5.32. These data suggests alcoholic extract of Black Peppers is almost as effective as Esb3 in coccidiosis.

Effect of Chutra on Oocyst Count

Oocyst count in group C (Chutra treated group) also showed significant reduction compared to control (group A) and Esb3 treated group (group D). Three chicks of group B and two chicks of group C died within 4-8 days of medication. Rest one and two chicks of group B and C respectively cured. It was observed that, 3 days after treatment, oocyst count started to fall unlike control group. On day 0, oocyst count in control group, Chutra treated group and Esb3 treated groups were 5272.50±31.98, 5265.00±57.81 and 5285±95.96 respectively. However, on day 3 during treatment, oocyst count reduced 54% in Chutra treated group and 62% in Esb3 treated group. On day 9, oocyst count in Chutra treated group and Esb3 treated group was 580 and 210±11.09 respectively compared to 6127 in control group. The number of oocyst count 6 days after post treatment was 37 in group B compared to 14±5.32 in group D (Table 1).

Table 1. Comparative efficacy of alcoholic extracts of Black Peppers, Chutra and Esb3 on oocyst count (per gm of feces) against experimentally induced coccidiosis in chickens

Group	Oocyst count during treatment and post treatment period						
	Treatment period					Post treatment period	
	Day 0	Day3	Day 6	Day 9	Day 12	Day 3	Day6
A	5272.50±31.98	5442.40±55.88	5945b	6127c	-	-	-
B	5212.40±94.20	3882.50±69.39	1652.30±36.83	580b*	130b	75b	37b
C	5265.00±57.81	2422.50±39.66	1575 ± 22.55a	450±29.26*a	125±17.32a	70±14.56a	32±12.20
D	5285±95.96	2032.5±87.97	615±124.09	210±11.09	52±8.62	24±7.39	14±5.32

The values above are mean ± SE of 4 chicks unless otherwise stated. a= values are mean ± SE of 3 chicks, b= values are mean of 2 chicks, c= value of 1 chick, * Significant decrease (p<0.01), - Death of all chickens.

Effect of Black Peppers, Chutra Leaves and Esb3 on Biochemical Parameters

SGPT level was calculated up to day 12 during treatment and up to day 6 post treatment. The data are presented in Table 2. On day 0, SGPT level was recorded in group A, B, C and D as 10.55±0.28, 9.425±0.30, 7.595±0.25 and 8.695±0.45 respectively. However, changes of SGPT level in all treatment groups were changes slightly and in significant in comparison to control group. However, the SGPT level was slightly increased from the initial value in group B, C and D and decreased in control group.

Table 2. Effect of alcoholic extracts of leaves of whole of Black Peppers, Chutra and patent drug ESB3 on SGPT (U/L) in chickens

Group	SGPT level on different days (U/L) of treatment and post treatment period						
	Treatment period					Post treatment period	
	Day 0	Day3	Day 6	Day 9	Day 12	Day 3	Day6
A	10.55±0.28	10.625±0.28	10.50b	10.5c	-	-	-
B	9.425±0.30	9.375±0.29	9.40a±0.46	9.60b±	9.50b0.70b	9.572b±0.715b	9.59b±0.15b
C	7.595±0.25	7.875±0.23	8.475±0.225a	8.40a±0.21a	8.23a±0.09a	8.40a±0.13a	8.42a±0.16a
D	8.695±0.45	8.65±0.39	8.725±0.36	8.725±0.49	8.825±0.30	8.765±0.36	8.83±0.39

The values above are mean ± SE of 4 chicks unless otherwise stated. a= values are mean ± SE of 3 chicks. b= values are mean of 2 chicks, c= value of 1 chick, *Significant decrease (p<0.01). -Death of all chickens

Effect of Black Peppers, Chutra Leaves and Esb3 on Blood Glucose Level

Blood glucose (U/L) level was calculated up to day 12 during treatment and up to day 6 post treatment in control group as well as treated groups. The data are presented in Table 3. No significant difference was observed in treatment group in comparison to control.

Table 3. Effect of alcoholic extracts of Black Peppers, Chutra and patent drug Esb3 on blood glucose (U/L) in chicken

Group	Glucose level on different days (U/L) of treatment and post treatment period						
	Treatment period					Post treatment period	
	Day 0	Day3	Day 6	Day 9	Day 12	Day 3	Day6
A	84.55±0.34	84.65±0.36	84.85b	85.4c	-	-	-
B	84.325±0.70	84.4±0.89	84.1a±0.89	84.95b	85.1b	84.94b	85.24b
C	84.65 ± 0.31	84.825 ± 0.34	84.825 ±0.36a	84.13 ± 0.12a	84.23a ± 0.12a	84.72 ± 0.23a	84.85a ± 0.24a
D	83.375±0.84	83.475±0.81	83.70±0.72	83.675±0.70	83.925±0.76	83.85±0.65	84.12±0.73

The values above are mean ± SE of 4 chicks unless otherwise stated. a= values are mean ± SE of 3 chicks, b= values are mean of 2 chicks, c= value of 1 chick, * Significant decrease (p<0.01), -Death of all chickens

Effect of Black Peppers, Chutra Leaves and Esb3 on Blood Parameters (TEC) Level in Chicken

Blood parameters (TEC) were observed in control and treatment groups. The data are presented in Table 4. No significant difference was observed in treatment group in comparison to control.

Table 4. Effect of alcoholic extracts of leaves of Black Peppers, Chutra leaves and patent drug Esb3 on blood parameter (TEC) in chicken

Group	TEC level on different days of treatment and post treatment period in chicken						
	Treatment period					Post treatment period	
	Day 0	Day3	Day 6	Day 9	Day 12	Day 3	Day6
A	1.425 ±	1.45a ± 0.17	1.46b	1.52c	-	-	-
B	1.85 ± 0.11	1.93 ± 0.062	2.25a ± 0.085	2.84b	3.16b	3.24b	3.18b
C	2.14 ± 0.04	2.35 ± 0.028	2.84 ± 0.066a	2.95 ± 0.1a	2.94 ± 0.082a	2.95 ± 0.067a	2.96a ± 0.058a
D	1.96 ±0.07	2.14±0.062	2.47±0.086	2.47±0.085	2.63±0.11	3.15±0.073	3.23±0.09

The values above are mean ± SE of 4 chicks unless otherwise stated. a= values are mean ± SE of 3 chicks, b= values are mean of 2 chicks, c= value of 1 chick, * significant decrease (p<0.01), Death of all chicks.

Effect of Black Peppers, Chutra Leaves and Esb3 on Blood Parameters (Hb) Level in Chicken

Blood parameter (Hb) was observed in control and treatment groups. The data are presented in Table 5. Hemoglobin level was significantly increased in Black Peppers treated group (from 7.32 ± 0.19 on day 0 of treatment period to 11.29 on day 6 post treatment period) compared to control (from 6.43 ± 0.13 on day 0 to 6.86 on day 9 before they died). The level of hemoglobin was also significantly increased (from 6.85 ± 0.13 on day 0 to 9.55 ± 0.53 on day 6 post treatment) in comparison to control. However, the Hb level obtained in group B is higher and group C is lower than Esb3 treated group (group D).

Table 5. Effect of alcoholic extracts of Black Peppers, Chutra leaves and patent drug ESB3 on blood parameter (Hb) in chicken

<table>
<tr><th rowspan="3">Group</th><th colspan="7">Haemoglobin level on different days (gm%) of treatment and post treatment period</th></tr>
<tr><th colspan="5">Treatment period</th><th colspan="2">Post treatment period</th></tr>
<tr><th>Day 0</th><th>Day3</th><th>Day 6</th><th>Day 9</th><th>Day 12</th><th>Day 3</th><th>Day6</th></tr>
<tr><td>A</td><td>6.43 ± 0.13</td><td>6.40 ± 0.12</td><td>6.40b</td><td>6.86c</td><td></td><td></td><td></td></tr>
<tr><td>B</td><td>7.32 ± 0.19</td><td>7.59 ± 0.37</td><td>8.20a ± 0.45</td><td>9.13b*</td><td>10.50b</td><td>10.80b</td><td>11.29b*</td></tr>
<tr><td>C</td><td>6.85 ± 0.13</td><td>7.30 ± 0.63</td><td>7.85a ± 0.23a</td><td>7.86a ± 0.13*a</td><td>8.35a ± 0.57a</td><td>8.73a ± 0.13a</td><td>9.55a ± 0.53a</td></tr>
<tr><td>D</td><td>7.57 ±0.17</td><td>8.13±0.63</td><td>8.85±0.23</td><td>9.29±0.53</td><td>9.78±0.13</td><td>10.16±0.62</td><td>10.57±0.13</td></tr>
</table>

The values above are mean ± SE of 4 chicks unless otherwise stated. a= values are mean ± SE of 3 chicks, b= values are mean of 2 chicks, c= value of 1 chick, * significant decrease (p<0.01), Death of all chicks

Effect of Black Peppers, Chutra leaves and Esb3 on PCV (%) in chicken

PCV (%) level was calculated up to day 12 during treatment and up to day 6 post treatment. The data are presented in Table 6. On day 0, SGPT level was recorded in group A, B, C and D as 28.53 ±0.13, 29.56±0.32, 28.36±0.36 and 27.56±0.25 respectively. However, PCV (%) in group B was significantly increased (31.82 on day 6 post treatment) in comparison to control and Esb3 treated group (30.03±0.73). PCV (%) in group C was also increased but significant in comparison to control and insignificant in comparison to Esb3 treated group.

Table 6. Effect of alcoholic extracts of Black Peppers, Chutra leaves and patent drug on PCV% in chicks

<table>
<tr><th rowspan="3">Group</th><th colspan="7">PCV level on different days (% In 30 min.) of treatment and post treatment period</th></tr>
<tr><th colspan="5">Treatment period</th><th colspan="2">Post treatment period</th></tr>
<tr><th>Day 0</th><th>Day3</th><th>Day 6</th><th>Day 9</th><th>Day 12</th><th>Day 3</th><th>Day6</th></tr>
<tr><td>A</td><td>28.53±0.13</td><td>28.23±0.12</td><td>27.56b</td><td>27.50c</td><td></td><td></td><td></td></tr>
<tr><td>B</td><td>29.56±0.32</td><td>29.15±0.15</td><td>38.50±0.62</td><td>30.58b*</td><td>31.16b</td><td>31.56b</td><td>31.82b*</td></tr>
<tr><td>C</td><td>28.36±0.36</td><td>28.78±0.16</td><td>29.26a±0.23a</td><td>29.67a±0.21a</td><td>29.83a±0.09a</td><td>30.26a±0.36a</td><td>30.79a±0.26a</td></tr>
<tr><td>D</td><td>27.56±0.25</td><td>28.120.16</td><td>28.56±0.23</td><td>29.23±0.56</td><td>29.56±0.21</td><td>30.83±0.56</td><td>30.03±0.73</td></tr>
</table>

The values above are mean ± SE of 4 chicks unless otherwise stated. a= values are mean ± SE of 3 chicks, b= values are mean of 2 chicks, c= value of 1 chick, * significant decrease (p<0.01), Death of all chicks

Coccidiosis is a ubiquitous intestinal protozoan infection of poultry seriously impairing the growth and feed utilization of infected bird. Conventional disease control strategies are now experiencing resistance resulting searching for alternative strategy to control the disease. There is a huge demand to investigate natural products to compete coccidiosis in poultry. The alcoholic extracts of whole of Black Peppers and leaves of Chutra at the dose rate of 9 ml/kg body wt. were given orally to group B and group C chicks respectively. Two chicks of group B died within 4-7 days of medication. Rest two chicks became apparently normal within 7-10 days of treatment. Oocyst disappeared completely 6 day onward of medication. Administration of alcoholic extracts of leaves of Chutra at the dose rate of 9 ml/kg body wt.in group C chicks gave 75% protection. In this group one chick died within 6 days of medication and rest three chicks survived. Oocyst disappeared completely from the faecal sample from 3rd day onward of medication. On the other hand, administration of Esb3 at the dose rate of 1 gm / litre of water afforded 100% protection in group D chicks. The oocyst disappeared completely from 3rd day onward of medication.

A number of medicinal plants and herbal preparation have been successfully used to control coccidiosis (Du and Hu, 2004; Dkhil et al., 2011; Almeida et al., 2014). Black pepper have been successfully used to reduced calcium level (Yoon et al., 2015), antimicrobials (Nassan & Mohamed, 2014), antioxidant (Oboh et al., 2013), anti-obesity (Neyrinck et al., 2013) and anti-inflammatory (Ying et al., 2013). However, no data of Black Pepper is yet available used it in coccidiosis. Therefore, the data could not be compared. But, it is observed that anti-inflammatory, anti-oxidant and antimicrobial properties could play a vital role as anti-coccidiosis. Chutra leaves have been shown effective as antidiabetic, hypolipidemic, and liver and renal damage recovering effects (Abedi Gaballu et al., 2015). Our data with Chutra leaves could not be compared as any data with coccidiosis treatment is available in published literature. However, this leaves are used as antidepressant and memory boosting (Patel & Udayabanu, 2013), hypoglycaemia (Patel & Udayabanu, 2013), anti-inflammatory (Hajhashemi & Klooshani, 2013) and antibacterial study (Körpe et al., 2013).

Therefore, it can be concluded that alcoholic extracts of Chutra leaves at the dose rate of 9 ml/kg was found to be most effective (75%) nearer to patent drug Esb3 in treating the chicks experimentally infected with coccidiosis. Alcoholic extracts of whole of Black Peppers proved less effective (50%) than those of alcoholic extracts of chutra leaves and patent drug Esb3. Further study should commence to investigate the mechanism of action of Black Peppers and Chutra leaves as anticoccidial in poultry.

REFERENCES

1. Abedi G, Fereydoon, YA Gaballu, OM Khyavy, A Mardomi, K Ghahremanzadeh, B Shokouhi, and H Mamandy, 2015. Effects of a Triplex Mixture of Peganum Harmala, Rhus Coriaria, and Urtica Dioica Aqueous Extracts on Metabolic and Histological Parameters in Diabetic Rats. Pharmaceutical Biology, 23:1-6.
2. Abou-Elkhair R, HA Ahmed, and S Selim, 2014. Effects of Black Pepper (piper Nigrum), Turmeric Powder (curcuma Longa) and Coriander Seeds (coriandrum Sativum) and Their Combinations as Feed Additives on Growth Performance, Carcass Traits, Some Blood Parameters and Humoral Immune Response of Broiler. Asian-Australasian Journal of Animal Sciences, 27: 847-54.
3. Almeida, Gustavo FD, Stig M Thamsborg, AMBN Madeira, JFS Ferreira, PM Magalhães, LC Dematté Filho, K Horsted, and JE Hermansen, 2014. The Effects of Combining Artemisia Annua and Curcuma Longa Ethanolic Extracts in Broilers Challenged with Infective Oocysts of Eimeria Acervulina and E. Maxima. Parasitology, 141: 347-55.
4. Coffin DL, 1955. Manual of Veterinary Clinical Pathology. 3rd Edn. Comstock Publishing Association, Inc, Ithaca, New York.
5. Conway DP and Mckenzie ME, 2007. Poultry Coccidiosis: Diagnostic and Testing Procedures. 3rd ed. Am. Blackwell Publishing Professional.
6. Dalloul, Rami A and Hyun S Lillehoj, 2005. Recent Advances in Immunomodulation and Vaccination Strategies against Coccidiosis. Avian Diseases, 49: 1-8.
7. Dkhil MA, AS Abdel-Baki, F Wunderlich, H Sies, and S Al-Quraishy, 2011. Anticoccidial and Antiinflammatory Activity of Garlic in Murine Eimeria Papillata Infections. Veterinary Parasitology, 175: 66-72.

8. Du A and S Hu, 2004. Effects of a Herbal Complex against Eimeria Tenella Infection in Chickens. Journal of Veterinary Medicine, Infectious Diseases and Veterinary Public Health, 51: 194-97.

9. Hajhashemi, Valiollah, and Vahid Klooshani, 2013. Antinociceptive and Anti-Inflammatory Effects of Urtica Dioica Leaf Extract in Animal Models. Avicenna Journal of Phytomedicine, 3: 193–200.

10. Harfoush MA, AM Hegazy, AH Soliman and S Amer, 2010. Drug Resistance Evaluation of Some Commonly Used Anti-Coccidial Drugs in Broiler Chickens. Journal of the Egyptian Society of Parasitology, 40: 337-48.

11. Körpe, D Aksoy, ÖD y İşerı, FI Sahin, E Cabi and M Haberal, 2013. High-Antibacterial Activity of Urtica Spp. Seed Extracts on Food and Plant Pathogenic Bacteria. International Journal of Food Sciences and Nutrition, 64: 355–62.

12. Nassan MA and EH Mohamed, 2014. Immunopathological and Antimicrobial Effect of Black Pepper, Ginger and Thyme Extracts on Experimental Model of Acute Hematogenous Pyelonephritis in Albino Rats. International Journal of Immunopathology and Pharmacology, 27: 531-41.

13. Neyrinck, Audrey M, M Alligier, PB Memvanga, E Névraumont, Y Larondelle, V Préat, PD Cani, and NM Delzenne, 2013. Curcuma Longa Extract Associated with White Pepper Lessens High Fat Diet-Induced Inflammation in Subcutaneous Adipose Tissue. PloS One, 8: e81252.

14. Oboh, Ganiyu, Ayokunle O Ademosun, Oluwatoyin V Odubanjo, and Ifeoluwa A Akinbola, 2013. Antioxidative Properties and Inhibition of Key Enzymes Relevant to Type-2 Diabetes and Hypertension by Essential Oils from Black Pepper. Advances in Pharmacological Sciences: 926047.

15. Patel, S Sharan and M Udayabanu, 2013. Effect of Urtica Dioica on Memory Dysfunction and Hypoalgesia in an Experimental Model of Diabetic Neuropathy. Neuroscience Letters, 552: 114–19.

16. Peek, HW and WJM Landman, 2011. Coccidiosis in Poultry: Anticoccidial Products, Vaccines and Other Prevention Strategies. Veterinary Quarterly, 31: 143-161.

17. Penev PM and L Lozanov, 1983. Action of Sodium Sulfachlorpyrazine (ESB3) on the Endogenous Development of Eimeria Tenella in the Experimental Infestation of Chickens. Veterinarno-Meditsinski Nauki, 20: 72-79.

18. Tewari AK, and BR Maharana, 2011. Control of Poultry Coccidiosis: Changing Trends. Journal of Parasitic Diseases, 35: 10-17

19. Xiaozhou Y, X Chen, S Cheng, Y Shen, L Peng, and HZ Xu, 2013. Piperine Inhibits IL-B Induced Expression of Inflammatory Mediators in Human Osteoarthritis Chondrocyte. International Immunopharmacology, 17: 293-99.

20. Yoon YC , SH Kim, MJ Kim, HJ Yang, MR Rhyu, and JH Park, 2015. Piperine, a Component of Black Pepper, Decreases Eugenol-Induced cAMP and Calcium Levels in Non-Chemosensory 3T3-L1 Cells. FEBS Open Biology, 5: 20-25.

3

EFFECT OF INCORPORATING RICE BRAN OR PARBOILED RICE POLISH WITH OR WITHOUT EXOGENOUS PHYTASE IN THE DIET ON GROWTH OF JINDING DUCKLINGS

Nebash Chandra Pal, Syed Mohammad Bulbul, Zannatul Mawa and Muslah Uddin Ahammad*

Department of Poultry Science, Faculty of Animal Husbandry, Bangladesh Agricultural University, Mymensingh-2202, Bangladesh

*Corresponding author: Muslah Uddin Ahammad; E-mail: ahammad.muslah@gmail.com

ARTICLE INFO	ABSTRACT

Key words

Rice bran
Parboiled rice polish
Phytase
Growth
Jinding duckling

A total of 48 straight-run day old ducklings (DOD) of Jinding were fed *ad libitum* on four (4) iso-nitrogenous and iso-caloric diets (3000 Kcal ME/kg and 22% CP) having 3 replicates each. The diets were formulated using 20% rice bran (RB) or parboiled rice polish (PRP) with or without exogenous phytase (10g/kg). Ducklings were fed up to 28 days of age to observe the effect of RB and PRP based diet on the growth performance. Feeding of PRP based diet with enzyme resulted in increased live weight gain and feed efficiency. There was no difference in feed intake on both diets ($p > 0.05$). However, the formulating cost of PRP based diet was higher than that of RB based diet. Addition of enzyme promoted growth and feed efficiency but did not affect feed intake significantly during the experimental period and increased feed cost. Therefore, it appeared that the biological performance of PRP based diet with or without enzyme was better than that of RB based diet. However, the feed cost was less in RB based diet with or without phytase. Therefore, to formulate low cost diet, RB seemed to be superior over PRP, but for better growth performance PRP can be used in the diet of ducklings. In conclusion, the findings demonstrated that the PRP based diets with or without phytase was superior to RB based diets with or without phytase in terms of growth performance, but RB based diet was superior in terms of feed cost. Therefore, it appeared that increased growth performance cannot be a basis of using RB and PRP. Rather, cost-effectiveness of feeding ducklings on RB and PRP based diet must be taken into account.

INTRODUCTION

Duck rearing is an integral part of rural poultry husbandry in Bangladesh. It provides with additional income especially for the rural women and acts as an important tool for poverty alleviation. According to the reports by FAO (2009), total duck population in Bangladesh is more than 18 million. Bangladesh is a riverine country. About 16488 km^2 of its total area are haors, baors, canals, ponds and low lying water reservoirs (Asian Livestock, 1978). The climatic conditions and innumerable water bodies of Bangladesh are suitable for duck habitation and production. Duck requires no additional care, management and supply of feed. They are hardier, capable to withstand the abuses of harsh climate with minimal management and inadequate nutrition, and are less susceptible to diseases compared with chicken. In addition, they are not competitor of chicken for feed because, they mainly scavenge in low-lying water lodged areas that are not suitable for chicken to scavenge around. It is well known that the indigenous (deshi) ducks can lay more eggs with larger size than the deshi chicken. Thus, ducks, being the most important versatile poultry species, can contribute to the increment in egg and meat production in Bangladesh.

In poultry production, it is well recognized that the feed cost alone accounts for about 65-70% of the total cost of production (Banerjee, 1992). In poultry farming of Bangladesh, high price and poor quality of most abundantly used feed ingredients are the major prevailing constraints, which may be alleviated through exploration of locally available cheaper potential feedstuffs and by introducing nutritional advanced technology to ensure high bio-availability of nutrients from the diets. In fact, poultry competes directly with human and other livestock for major grains such as wheat and maize, which are usually incorporated at a rate of 50-60% in the total diet formulated for feeding poultry. However, they are less available as compared to their demand. Therefore, it is imperative to explore the possibilities of using locally available cheaper grains or nutritionally equivalent their by-products to reduce feed cost. RB and PRP are the only by-products of abundantly grown grains, the paddy in Bangladesh. In relevance, when compared with chicken, ducks have been shown to be capable of consuming and utilizing more fibrous diet very efficiently. RB is a by-product of paddy obtained from rice milling processes. It consists of the combined aleurone and pericarp (McDonald, 1987).

In Bangladesh, most of the farmers, who rear ducks traditionally, use RB for the feeding of ducks. Since RB contains more fiber of coarse type, it is not well utilized by physiological process of ducks which ultimately reduces growth performance of ducks. Juliano (1985) reported that RB also contains anti-nutritional factors. RB has higher phytin P content (24 to 46 g/kg) than other cereal brans (Thomson and Weber, 1981; Warren et al., 1990) which is relatively unutilizable to poultry. A heat stable factor pepsin inhibitor, present in both bran and germ was identified as phytin P (Kanaya et al., 1976). However, the duck producers in Bangladesh do not know about the bad effect of using RB as duck feed. In contrast, the other potential by-product of paddy is rice polish (RP), which has been shown to contain only 2.7% fiber and as much protein and fat as in RB (Morrison, 1957). It has been reported that RP constitutes about 10% of paddy and is available in large quantities in major paddy growing areas of the world (Houston and Kohler, 1970). It might be used as an alternative to grains and RB in feeding ducks. There are 3 major types of RP namely raw RP, deoiled RP and PRP. The raw RP is not a suitable feed for poultry feeding. It has been shown that PRP is more stable to the oxidative hydrolysis and is less susceptible to the development of free fatty acids to be rapidly oxidized during storage than that of raw RP (Shaheen et al., 1975). Due to application of modern techniques in agriculture, paddy production in the country has increased. During milling in hullers considerable amount of breakage of rice occurs which reduces the quantity of consumable rice of a given quantity of paddy. To withstand the milling pressure, rice kernel has to be hardened by some preprocessing techniques. Partial cooking of grain with intact husk to impart desirable hardness of kernel is termed as parboiling. Polish obtained from parboiled paddy in automatic rice mills is called PRP. Chemical analysis revealed that the PRP contains 13% CP and 3250 kcal ME/kg which is very much comparable to the nutrient content of wheat and maize (Zablan et al., 1963; Scott et al., 1976; Shivaji et al., 1983 and Eshwaraiah et al., 1988). However, PRP contains phytate phosphorus (Padua and Juliano, 1974; Barber and Barber, 1980), which reduces the phosphorus and calcium availability responsible for depressed performance of poultry on PRP diet. PRP also contains some non-starch polysaccharides such as cellulose, xylose, arabinose and galactonic acid that are not easily digested by poultry. The anti-nutritional effect is manifested by poor growth accompanied by depressed nutrient utilization. Recently, nutritionists are trying to utilize the efficiency of fibrous diet using different additives including enzymes. Addition of exogenous phytase and carbohydrase has been reported to improve feed utilization in broiler on PRP based diet (Moshad, 2001). Phytase that used in poultry diet also helps in reducing environmental pollution (Kies et al., 2001). Therefore, the adverse effects of PRP could possibly be overcome by dietary supplementation of exogenous phytase. Very few researchers examined the effect of using PRP fortified with phytase in the diet of ducks. Therefore, the present study was aimed at examining the effect of PRP and RB based diet with or without phytase on the growth performance of ducklings.

MATERIALS AND METHODS

The experiment was carried out in a shed type, well ventilated open-sided house for a period of 28 days. Allocated floor space was 900 cm^2 per duckling. The whole wire meshed floor area was divided into a total of 12 uniform pens made of bamboo and wire net. A total of 48 as hatched individually wing banded Jinding DOD were randomly allocated to four (4) different dietary treatment groups, which was replicated three (3) times each. The diets were prepared replacing grain(s) by RB or PRP at a level of 20% with or without enzyme. Ducklings were fed *ad libitum* throughout the experimental period on 4 different iso-nitrogenous and iso-caloric diets namely: D_1 (containing 20% RB without phytase), D_2 (containing 20% RB + 1% phytase), D_3 (containing 20% PRP without phytase) and D_4 (containing 20% PRP + 1% phytase). The nutrient levels of the four diets were adjusted in accordance with the feeding standard recommended by NRC (1994). The test diets were formulated using locally available feed ingredients including RB, PRP, maize, soybean meal, protein concentrate (Jasoprot), L-lysine, DL-methionine, di-calcium phosphate, vitamin-mineral-amino acid premix, common salt and phytase enzyme (Rena-Phytase-400; BASF, DSM, Renata Animal Health, Bangladesh). Feeds were supplied in mash form. One trough feeder and one round drinker were provided to each pen.

Table 1. Chemical composition of the experimental balanced ration

Nutrient	Amount
Metabolizable energy (Kcal ME/kg)	3000
Crude protein (%)	22.0
Crude Fiber (%)	6.0
Calcium (%)	0.34
Phosphorus (%)	0.65
Lysine (%)	1.14
Methionine (%)	0.36

During the whole experimental period, all ducklings were exposed to a continuous lighting of 16 hours including day light and artificial light. Artificial lighting was reduced by 1 hour/week. Individual records were kept on initial and weekly live weight, weekly feed intake, feed conversion ratio (FCR) and survivability. All data either measured or calculated regarding growth, feed intake, FCR and feed cost were analysed by one-way ANOVA with the generalized linear model (GLM) using IBM SPSS Statistics version 19 (SPSS Inc., an IBM Company, Chicago, IL). In some cases where analyses revealed a significant treatment effect, the differences between mean values were evaluated using LSD (Least Significant Difference). A value of $P < 0.05$ was considered statistically significant, unless stated otherwise.

RESULTS AND DISCUSSION

Live weight

Live weight difference among ducklings on different dietary treatments was negligible (p>0.05) at day old, but differences in live weight was marked (p<0.01) with the increase of their age. PRP based diet fortified with phytase showed a definite superiority in increasing live weight gain of ducklings irrespective of age, over RB based diet (p<0.01) (Table 2). Live weight of ducklings on PRP based diet without enzyme was increased by 7.15 % than that on RB based diet. On the other hand, live weight on PRP based diet with enzyme increased by 7.36 % than that on RB based diet. Superiority of PRP to RB irrespective of exogenous phytase, might be due to the removal of different toxins; saponin, pyridine, hemagluttinin and tannin during parboiling. Such type of possibility has been supported by Eshwaraiah et al., (1986). Islam (1994) reported that live weight decreased on diets containing increasing level of PRP. It decreased growth when levels of RP in the diet were above 40% (Chaturvedi and Mukherjee, 1967). Kamal (1993) carried out an experiment with Khaki Campbell ducks under village condition and observed that supplementation of RP resulted in increased body weight gain in comparison with those receiving no RP. Increased live weight on RB and PRP based diet with enzyme was in agreement with Naher (2002), Moshad (2001) and Seskeviciene et al., (1999). Naher (2002) showed that addition of mixed enzyme increased meat yield of ducks.

Feed intake

Feed intake did not differ (p>0.05) between PRP and RB based diets with or without exogenous phytase irrespective of age (Table 2), implying that the performance in terms of weight gain was influenced by diets with or without phytase without affecting feed intake. The result of adding coarse fibrous RB in the diets indicates that the volume of RB based diet may be an important factor to limit feed intake. Our results are in agreement with the findings of Steenfeld et al., (1998), who used cell wall degrading enzyme at the rate of 63 or 70 g/kg wheat based diet and four different enzyme preparations (2 xylanase preparations and 2 mixed enzyme preparations) to more than 80 g/kg of wheat based diet. In both cases, they observed that feed intake was not influenced by enzyme addition. Similar to the current findings, Richter et al., (1991) reported that enzyme supplementation has no impact on extra feed intake. However, Naher (2002) showed that addition of phytase and carbohydrase in the ration of ducks increased feed intake. In contrast to the present findings, Islam (1994) reported that feed intake increased with increasing PRP levels from 5.00 to 31.50% but decreased at 40.95% dietary PRP.

Table 2. The live weight, feed intake, feed conversion ratio and feed cost of ducklings on rice bran and parboiled rice polish based diets with or without enzyme at different ages

Parameter	Age (days)	RB (20%)		PRP (20%)		SED and Significance[+]
		Phytase (0g/kg)	Phytase (10g/kg)	Phytase (0g/kg)	Phytase (10g/kg)	
Live weight	0	49.67	50.33	51.67	49.33	1.130[NS]
	7	129.67[d]	136.00[c]	140.00[b]	147.00[a]	2.060**
	14	231.33[d]	242.33[c]	248.67[b]	261.67[a]	2.940**
	21	422.33[d]	440.00[c]	448.33[b]	465.00[a]	3.590**
	28	508.00[d]	530.00[c]	544.33[b]	569.00[a]	4.460**
Feed intake	7	127.67	124.33	127.00	126.00	2.590[NS]
	14	210.33	210.67	211.00	210.33	2.660[NS]
	21	311.00	311.67	311.00	310.00	1.260[NS]
	28	387.67	389.33	390.33	389.00	2.980[NS]
Feed conversion ratio	7	1.60[a]	1.45[b]	1.44[b]	1.29[c]	0.030**
	14	2.07[a]	1.98[b]	1.94[b]	1.82[c]	0.030**
	21	1.64[a]	1.58[b]	1.56[bc]	1.53[c]	0.030**
	28	4.55[a]	4.33[b]	4.07[c]	3.75[d]	0.080**
Feed cost	7	2.55[d]	2.85[b]	2.66[c]	3.02[a]	0.060**
	14	4.21[b]	4.84[a]	4.43[ab]	4.72[ab]	0.340[NS]
	21	6.22[d]	7.16[b]	6.52[c]	7.45[a]	0.030**
	28	7.75[d]	8.95[b]	8.20[c]	9.33[a]	0.070**

[+] NS, p>0.05; **, p<0.01; RB = rice bran; PRP = parboiled rice polish; SED = standard error difference.

Feed Conversion Ratio

FCR was higher in RB based diet than that on PRP based diet. FCR on PRP based diet without enzyme was improved by 10.55 % than that on RB based diet. FCR on PRP based diet with enzyme was improved by 13.39 % than that on RB based diet. The recorded lower FCR on PRP based diets with or without enzyme than that on RB based diets coincided with the findings of Moshad (2001), who reported improved feed conversion on PRP based diet with the addition of enzyme. Phytase that are used in poultry diet also helps in reducing environmental pollution (Kies et al., 2001). Phytin phosphorus is located in globoides in the aleurone protein bodies as potassium, magnesium salts. Its phosphate group can readily form complex with calcium, zinc, iron, protein and starch. Phytase can liberate (a proportion of) these compounds, thereby increasing the energy and protein value of the diet (Kies et al., 2001).

Feed cost

Feed cost on PRP based diet without enzyme was increased by 5.8 % than that on RB based diet. Feed cost on PRP based diet with enzyme was increased by 4.25 % than that on RB based diet. Higher feed cost on PRP based diets with enzyme supplementation is not supported by the findings of Farrell et al., (1993). The increasing performance on PRP diet in terms of live weight and feed conversion has been counteracted by remarkable decrease in feed cost on RB diet. The findings of the present study coincided with the results reported by Mikulshi et al., (1999), who replaced all wheat, two-third maize by triticale and barley supplemented with enzyme preparation (contained beta-glucanase, cellulase, protease and amylase) in their study. They showed that enzyme supplementation decreased the relative cost of feeds by 4-12% compared with the wheat-maize based control diet. It has also been stated that multi-enzyme supplementation to commercial broiler diet decreased feed cost by 8.81 to 9.73% for production of 1 kg broiler meat (Augelovicova and Michalik, 1997).

Behaviour of ducklings

Table 3. The behaviour; Proportionate time spent on feeding (Fd), drinking (Dk), standing (St), and resting (Rt) of broilers fed on rice bran (RB) and parboiled rice polish (PRP) based diets

Parameter	RB (200g/kg)		PRP (200g/kg)		SED and significance [+]
	Phytase (0g/kg)	Phytase (10g/kg)	Phytase (0g/kg)	Phytase (10g/kg)	
Fd.	16.55	17.33	17.67	18.00	1.620[NS]
Dk.	14.67	14.83	15.17	14.50	1.170[NS]
St.	47.50[a]	47.33[b]	40.33[a]	40.00[b]	2.600**
Rt.	20.00[b]	20.50[b]	26.50[a]	28.83[a]	2.200**

+ NS, p>0.05, **, p<0.01; All SEDs are against 3 error degrees of freedom

Regardless of phytase supplementation, proportionate time spent on feed intake was similar between RB and PRP based diets (Table 3). It appears that time spent on drinking seemed to be a simple function of feed intake. The birds having higher feed intake perhaps spent similar time in standing. Ducklings on PRP based diets with or without phytase took more rest than that on RB based diets with or without exogenous phytase. Taking more time for standing and resting indicates better feed conversion efficiency. The ducklings took more time for standing during feeding on RB based diet with or without enzyme than that on PRP based diets with or without enzyme. The reason behind this occurrence might be due to the increased volume of RB in the diets. Due to increased volume of RB in diets RB with or without enzyme, the birds took more time in feeding. As a result, birds had to stand more time for feeding. The behavioural pattern, irrespective of feeding, drinking, standing, and resting suggests that the behavioural component, especially time spent in resting and consequent feed conversion showed a significant positive relation. Appreciable increase in feed conversion efficiency obtained against higher resting time implies that time spent in activity incurs remarkable need of energy in feeding resulted efficient feed utilization. But the current information for such an occurrence could not be compared for the paucity of the published information at the present time.

CONCLUSION

Based on the findings of the current study, it may be concluded that the inclusion of costlier PRP in the diet of ducklings resulted in higher growth rate and better feed efficiency as compared to that of cheaper RB. Supplementation of enzyme accelerated feed utilization as indicated by the increased performances of ducklings on diet with enzymes. Therefore, it is suggested to use phytase-fortified PRP based diet for feeding ducks. Augmentation of production performance seems to be a better option for the duck producers than the reduction of feed cost.

REFERENCES

1. Asian Livestock, 1978. FAO. Regional Animal Production and Health Commission for Asia and the Pacific (APHCA). 10: 6.
2. Augelovicam M and I Michalik, 1997. A test of enzymatic preparation in relation to performance and commercial utilization of feeds in broiler chickens. 42: 175-180.
3. Banerjee G C, 1992. Classification and composition table for poultry feeds. 3rd edition, 105-108.
4. Barber S K and P K Barber, 1980. Utilization of rice bran by broiler and layer chicken. Poultry Science, 59: 1012-1017.
5. Chaturvedi, D. K. and Mukherjee, R. 1967. Studies on cereal free rations based on rice polishing and groundnut cake for growing chicks. Indian Journal of Poultry Science, 2: 36-51.
6. Eshwaraiah; Reddy C V and V P Rao, 1988. Effect of autocalaving and solid substrate fermentation of raw, deoiled and parboiled rice polishing in broiler diets. Indian Journal of Animal Science, 58: 377-381.
7. FAO, 2009. Food and Agriculture Organization. Year Book, 57: 211.
8. Houston D F and G O Kohler, 1970. Nutritional of rice. National Academy of Science. Washington. D.C.
9. Islam NM, 1994. Parboiled rice polish as a dietary substitute of wheat on growth performance and meat yield of broilers. Ph.D. Thesis. Department of Poultry Science, Bangladesh Agricultural University, Mymensingh, Bangladesh.
10. JulianoBO (editor), 1985. Rice Chemistry and technology 2nd edition American Association of cereal chemists Inc. St. Paul, Minnesota, U.S.A.
11. Kamal MM, 1993. Study on the effect of supplementation of rice polish on the performance of Khaki Campbell Duck under village condition. M.Sc. Thesis, Department of Animal Nutrition, Bangladesh Agricultural University, Mymensingh, Bangladesh.
12. Kanaya K,Yasumoto K and H Mitsuda, 1976. Pepsin inhibition by phytate contained in rice bran, In: B. O. Juliano (editor) 1985. Rice chemistry and technology 2nd edition. American Association of Cereal Chemists. Inc. St. Paul. Minnesota, USA.
13. Kies A K,Vanhemert K A F and W C Saur, 2001. Effect of phytase on protein and amino acid acid digestibility and energy utilization. World's Poultry Science Journal, 57: 109-124
14. McDonald P, Edwards R A and J F D Greenhulgh, 1987. Nutritive value of food for poultry. Animal Nutrition 4th edition, Longman Group Ltd. U.K.
15. Mikulski, D.; Jankowski, J.; Farugh, A. and Zeid, A. E. 1999. Effect of feeding enzyme-supplemented triticale-bareley diets on broiler chicken. Egyptian poultry Science Journal, 19: 607-618.
16. Morrison F B, 1957. Feed and feeding 2nd edition. Marrison Publishing Co. Ithaca, New York.
17. Moshad M A, 2001. Use of phytase and carbohydrase enzyme for better utilization of parboiled rice polish based diet in Broilers. M.S. Thesis, Department of Poultry Science, Bangladesh Agricultural University, Mymensingh.
18. Naher B, 2002. Utilization of parboiled rice polish based diet with supplementation of carbohydrase and phytase in growing ducklings. M. S. thesis, Department of Poultry Science, Bangladesh Agricultural University, Mymensingh.
19. Pauda I B and C H Julinao, 1974. Phytase on the utilization of chicks. Journal of nutrition, 104: 203-2038.

20. Richter G, Lemsef A, Cyriaci G and J Schwartze, 1991. Evaluation of microbial phytase in broiler feeding. Poultry Abstracts, 1993. 19.

21. Scott M L,Nesheim M C and R J Young, 1976. Nutrition of the chicken, 2nd edition. M. L. Scott Associates. Ithaca, New York.

22. Shaheen A B and Shirbeeny, 1975. Effect of parboiling of rice on the rate of lipid hydrolsis and deterioration of rice bran. Cereal chemistry, 52: 1-8.

23. Shivaj S, Zambade S S,Jagmohan S and J S Ichnopnani, 1983. Nutritive value of raw, parboiled, Stabilized and deoiled rice bran for growing chicks. Journal of Food and Agriculture, 34: 743-788.

24. Steenfeld S,Mulleriz A and J F Jenson, 1998. Enzyme supplementation of Wheat based diets for broilers. Effect on growth performance and intestinal viscosity. Animal Feed Science and Technology, 75: 27-43.

25. Thompson S A and CW Weber, 1981. Effect of dietary fibre sources on tissue mineral level in chicks. Poultry Science, 60: 840-845.

26. Warren B F and D J Farrell, 1990. The nutritive value of full-fat and defatted Austrialian rice bran. I. Chemical Composition. II Growth studies with chicken, rats and pigs. Animal Feed Science and Technology, 27: 229-246.

27. Zablan T A, Griffith A M, Nesheim M C, Young R J and ML Scott, 1963. Matabolizable energy of some oil seed meals and unusual feedstuffs. Poultry Science, 42: 619-625.

4

IDENTIFICATION OF BACTERIAL AGENTS FROM THE FAECAL SAMPLES OF DIARRHOEIC SHEEP AND THEIR ANTIBIOTIC SENSITIVITY

Md. Nuruzzaman Munsi*, Md. Ershaduzzaman, Md. Osman Gani, Md. Moinuddin Khanduker and Md. Shahin Alam

Goat and Sheep Production Research Division, Bangladesh Livestock Research Institute, Savar, Dhaka-1341, Bangladesh

*Corresponding author: Md. Nuruzzaman Munsi; E-mail: nzaman_blri@yahoo.com

ARTICLE INFO	ABSTRACT

Key words

Bacteria
Diarrhoea
Sheep
Antibiotic
Sensitivity

The current study was carried out to identify the bacterial species in the faecal samples of 20 diarrhoeic sheep and to observe their sensitivity to different antibiotics. This investigation was performed by collecting diarrhoeal samples from the sheep (n=20) under goat and sheep research farm of Bangladesh Livestock Research Institute (BLRI), Savar, Dhaka, Bangladesh. The average age and body weight of the animals were 25 days and 2.5 kg respectively. Of the bacteria responsible for diarrhoea in case of sheep, *Escherichia coli* alone was found in 6 samples (30%), *Escherichia coli* combined with *Proteus mirabilis* was found in 12 samples (60%), and no bacteria could be detected in 2 of the 20 samples tested. Both the bacteria were highly sensitive to ceftriaxone, and moderately sensitive to ciprofloxacin and gentamicin. It could be stated that ceftriaxone is the antibiotic of first choice for the treatment of diarrhoea in sheep, where *Escherichia coli* is suspected to be the principal causal agent of diarrhoea.

INTRODUCTION

Diarrhoea characterized by an increased frequency, fluidity, or volume of faecal excretion is a major problem in the farm animals. It is caused by a wide range of microbial, parasitic and environmental factors namely–bacteria, virus, helminths, protozoa, toxic substances, sands, lush pasture, overfeeding, overcrowding, poor sanitation, inadequate intake of colostrum, poor quality colostrum, poor quality milk replacers, feeds difficult to digest, etc. On the basis of literature the bacterial agents causing diarrhoea in sheep are *Escherichia coli*, *Salmonella spp.* and *Clostridium perfringens,* where *Proteus spp.*, and some other species of bacteria may be associated in some cases (Rahaman, 1995).

According to Hindson and Winter (2002), a combination of reduced reabsorption from the lumen of the intestine, and increased fluid loss through damaged mucosa into the intestine is found in many forms of both gastroenteritis and inflammatory change in the lower intestine. Such a condition ranges from bacterial infections such as *E. coli* and *Salmonella* to parasitic diseases such as coccidiosis and parasitic gastroenteritis. Diarrhoea or enteritis, to a greater extent, is related with clostridial diseases. These clostridial diseases of sheep have been recognized clinically for over 200 years, but not until the end of the nineteenth century did their bacterial nature start to be unraveled, a process that continued over the next 50 years (Sterne, 1981). Even during the 1990s, new information came to light as the importance of Clostridium sordellii as a cause of abomasitis and enteritis in all ages of sheep was established (Lewis and Naylor, 1998). The haemorrhagic enteritis affecting lambs in the first few days of life is caused by *C. perfringens* types (either beta or beta 2 toxin) or C, and differs from lamb dysentery only in the gross pathology and being marginally less acute, affecting lambs up to 3 weeks of age (Lewis, 2007).

E. coli is a major cause of diarrhoea in calves, piglets and lambs, and the term 'colibacillosis' is commonly used. It causes huge economic loss in this age group of animals (Radostits *et al.*, 2000). However, selection of suitable antibiotics is a good strategy for successful treatment of bacterial diarrhoea. But all antibiotics are not always useful for the treatment of diarrhoea because of development of antibiotic resistance. So, the present study was carried out to identify the bacterial species in the faecal samples of diarrhoeic sheep and to observe their sensitivity to different antibiotics.

MATERIALS AND METHODS

This investigation was performed in the bacteriological laboratory of Bangladesh Livestock Research Institute (BLRI) by collecting diarrhoeal samples from the sheep (n=20) in BLRI goat and sheep research farm. The average age and body weight of the animals were 25 days and 2.5 kg, respectively. The duration of the experiment was from August, 2010 to May, 2011.

Sample collection

The faecal samples were collected aseptically into stool containers directly from the rectum with the help of small polybags.

Laboratory tests for identification of bacteria

The collected samples were then allowed for culture in bacteriological media (nutrient agar, EMB agar, SS agar, Blood agar, Mac Conkey agar), Gram's staining and biochemical tests (Methyl Red test, Indole test, Catalase test, etc) for isolation and identification of the bacterial species.

Antibiotic sensitivity test

The antibiotic sensitivity test was done by using disc diffusion method (Cowan and Steel, 1965). The commercially available antibiotic discs such as, oxytetracycline, gentamicin, amoxycillin, penicillin G, cloxacillin, sulphamethoxazole, streptomycin, ciprofloxacin and ceftriaxone were used to know the sensitivity of *Escherichia coli* and *Proteus mirabilis* to these drugs.

RESULTS AND DISCUSSION

The isolates were identified as *Escherichia coli* on the basis of morphology (Gram negative rod), cultural characteristics (Green metallic sheen on EMB agar as in fig.1) and biochemical characteristics and *Proteus mirabilis* on the basis of morphology (Gram negative rod), cultural characteristics (Swarming growth on Mac Conkey agar as in fig.2 and Brilliant Green Agar) and biochemical characteristics. Among the different kinds of bacteria responsible for diarrhoea in sheep, only *Escherichia coli* was found in 6 samples (30%), *Escherichia coli* alone combined with *Proteus mirabilis* was found in 12 samples (60%), and no bacteria were found in 2 of the 20 samples tested (Table 1).

Both the bacteria were highly sensitive to ceftriaxone, and moderately sensitive to ciprofloxacin and gentamicin (Table 2). The inhibition zones, in case of ceftriaxone, were 29 mm for *Escherichia coli* and 27 mm for *Proteus mirabilis* while these zones were 17 mm for *Escherichia coli* and 21 mm for *Proteus mirabilis* in case of ciprofloxacin and 17 mm for both *Escherichia coli* and *Proteus mirabilis* in case of gentamicin (Table 2). These bacteria were found resistant to other antibiotics used in the sensitivity test (Table 2). Wang *et al.* 2014 noticed decreased susceptibility of *P. mirabilis* in Taiwan to some broad spectrum antibiotics, including 3rd-generation cephalosporins and ciprofloxacin, whereas Kwiecińska-Piróg *et al.* 2013 found that in most of the tested concentrations, ciprofloxacin was more efficient than ceftazidime against the *P. mirabilis* biofilm. They became able to prove that the efficiency of antibiotics against *P. mirabilis* biofilm depends on its maturity and strains' origin.

Bacterial resistance to a particular antibiotic might occur due to use of that antibiotic for a longer period of time. The present study revealed that *E. coli* isolates were resistant to several antibiotics like penicillin G, amoxycillin, cloxacillin, streptomycin and oxytetracycline. These findings are in support of Ershaduzzaman *et al.* 2007, Tadesse *et al.* 2012. But Islam *et al.* 2007 found streptomycin as a sensitive antibiotic to *E. coli* which contradicts the present findings. However, this might happen due to strain variation of *E. coli*.

In this investigation the highest sensitivity was recorded for ceftriaxone because it is a new generation of cephalosporin and has not been used by the physician for long time.

Table 1. Prevalance of bacteria found in the diarrhoeal samples of sheep

Bacteria Found	Total Number of Samples Tested	No. of Positive Samples	Percentage
Escherichia coli and *Proteus mirabilis.*	20	12	60 %
Escherichia a coli	20	6	30 %
No bacteria	20	2	10 %

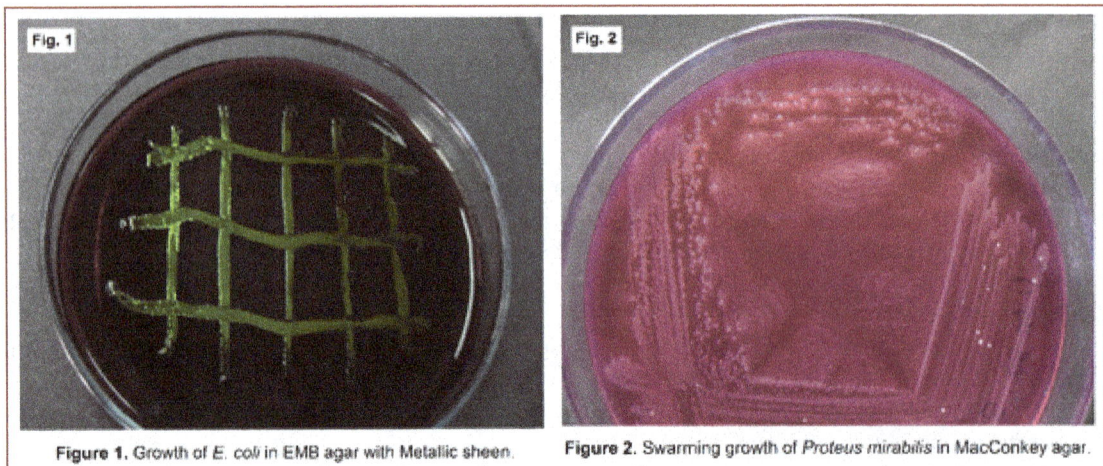

Figure 1. Growth of *E. coli* in EMB agar with Metallic sheen.

Figure 2. Swarming growth of *Proteus mirabilis* in MacConkey agar.

Table 2. Inhibition zones produced by antibiotics used against *Escherichia coli* and *Proteus mirabilis* in the sensitivity test.

Antibiotics discs used	Bacterial species	Zone of inhibition	Result
Oxytetracycline	*Escherichia coli*	7 mm	Resistant
	Proteus mirabilis	0 mm	Resistant
Gentamicin	*Escherichia coli*	17 mm	Moderately sensitive
	Proteus mirabilis	17 mm	Moderately sensitive
Amoxycillin	*Escherichia coli*	0 mm	Resistant
	Proteus mirabilis	7 mm	Resistant
Penicillin G	*Escherichia coli*	0 mm	Resistant
	Proteus mirabilis	0 mm	Resistant
Cloxacillin	*Escherichia coli*	0 mm	Resistant
	Proteus mirabilis	0 mm	Resistant
Sulphamethoxazole	*Escherichia coli*	0 mm	Resistant
	Proteus mirabilis	0 mm	Resistant
Streptomycin	*Escherichia coli*	0 mm	Resistant
	Proteus mirabilis	0 mm	Resistant
Ciprofloxacin	*Escherichia coli*	17 mm	Moderately sensitive
	Proteus mirabilis	21 mm	Highly sensitive
Ceftriaxone	*Escherichia coli*	29 mm	Highly sensitive
	Proteus mirabilis	27 mm	Highly sensitive

N.B.: The interpretation was done as resistant (≤10 mm), less sensitive (11-14 mm), moderately sensitive (15-18 mm) and highly sensitive (≥19 mm) except penicillin G, where zone of inhibition for resistance range is ≤28 mm according to Kirby-Bauer Method.

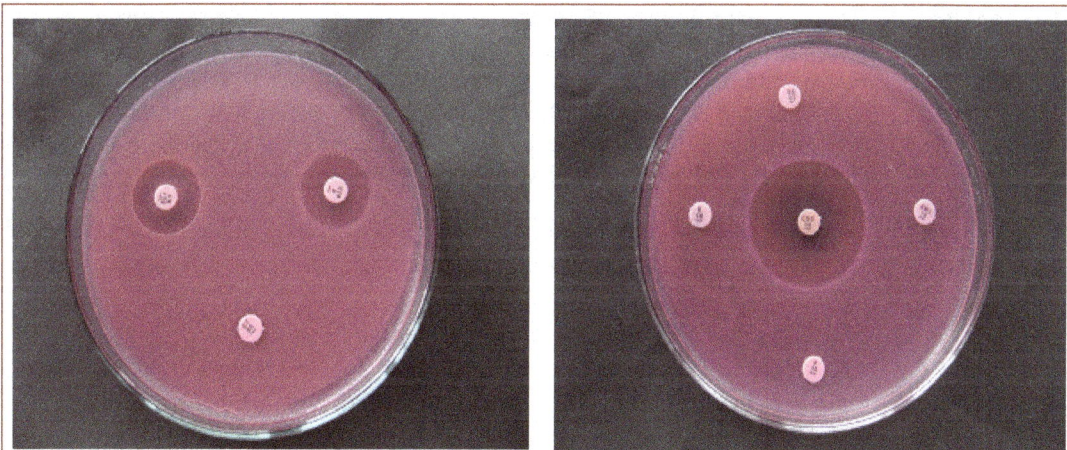

Figure 3. Production of clear zone of inhibition by Ceftriaxone (at the right-central), Ciprofloxacin (at the left-left one), Gentamicin (at the left-right one) of *E. coli in* Mac Conkey agar.

CONCLUSION

In conclusion, it might be stated that ceftriaxone is the antibiotic of first choice, and ciprofloxacin and gentamicin are the antibiotics of second choice to be used for the treatment of diarrhoea in sheep, where *Escherichia coli* is suspected to be principal causal agent of diarrhoea.

REFERENCES

1. Cowan ST and Steel KJ, 1965. Manual for the Identification of Medical Bacteria. Cambridge University Press, Cambridge.
2. Ershaduzzaman M, Taimur MJFA and Munsi MN, 2007. Bacteriopathology of Pneumoenteritis and Antibiotic Sensitivity of the Organisms Isolated from Black Bengal Kids Affected with the Diseases. Bangladesh Journal of Livestock Research, 14: 59-66.
3. Hindson JC and Winter Agnes C., 2002. Diarrhoea. In Manual of Sheep Diseases, 2nd edn., Blackwell Science Ltd., pp 90-95.
4. Islam MR, Ershaduzzaman M, Faruque MH, Munsi MN, Alam MS and Talukder MAI, 2007. Isolation and Identification of the Organisms of Sub-clinical Mastitis of Sheep and Goats in BLRI Sheep and Goat Farm. Proceedings of Annual Research Review Workshop-2007, Bangladesh Livestock Research Institute, Savar, Dhaka 1341, pp: 60-61.
5. Kwiecińska-Piróg J, Skowron K, Zniszczol K, and Gospodarek E, 2013. The Assessment of Proteus mirabilis Susceptibility to Ceftazidime and Ciprofloxacin and the Impact of These Antibiotics at Subinhibitory Concentrations on Proteus mirabilis Biofilms. BioMed Research International, Article ID 930876, doi.org/10.1155/2013/930876.
6. Lewis CJ, 2007. Clostridial Diseases. In Diseases of Sheep, 4th edn., Blackwell Publishing Ltd., pp 156-59.
7. Lewis CJ and Naylor RD, 1998. Sudden Death in Sheep Associated with *Clostridium sordellii*. Veterinary Record, 142: 417-21
8. Radostits OM, Gay CC, Blood DC and Hinchcliff KW, 2000. Diseases Caused by Bacteria – III. In Veterinary Medicine, 9th edn., pp 779-83.
9. Rahaman M, 1995. Systemic Bacteriology. In Bacteriology, 1st edn., p 134.
10. Sterne M, 1981. Clostridial Infections. British Veterinary Journal, 137:443-54.
11. Tadesse DA, Zhao S, Tong E, Ayers S, Singh A, Bartholomew MJ, and McDermott PF, 2012. Antimicrobial Drug Resistance in *Escherichia coli* from Humans and Food Animals, United States, 1950–2002. Emerging Infectious Diseases, 18(5): http://dx.doi.org/10.3201/eid1805.111153
12. Wang J-T, Chen P-C, Chang S-C, Shiau Y-R, Wang H-Y, Lai J-F, Huang I-W, Tan M-C, Lauderdale T-LY, and TSAR Hospitals, 2014. Antimicrobial susceptibilities of *Proteus mirabilis*: a longitudinal nationwide study from the Taiwan surveillance of antimicrobial resistance (TSAR) program. BMC Infectious Diseases, 14: 486.

5

CLINICAL PREVALENCE OF DISEASES AND DISORDERS OF CATTLE AT THE UPAZILLA VETERINARY HOSPITAL, CHAUHALI, SIRAJGANJ

SHM Faruk Siddiki[1*], Mohammad Golam Morshed[2], Mst. Sonia Parvin[3] and Lutfun Naher[3]

[1]Faculty of Veterinary Medicine and Animal Science, Bangabandhu Sheikh Mujibur Rahman Agricultural University, Gazipur, Bangladesh; [2]Upazilla Veterinary Hospital, Chauhali, Sirajganj, Bangladesh; [3]Department of Medicine, Faculty of Veterinary Science, Bangladesh Agricultural University, Mymensingh, Bangladesh

*Corresponding author: SHM Faruk Siddiki; E-mail: ufs.vet@gmail.com

ARTICLE INFO

ABSTRACT

Key words

Clinical prevalence
Diseases
Disorders
Cattle

An investigation was undertaken to determine the general clinical prevalence of diseases and disorders in cattle at the Upazilla Veterinary Hospital, Chauhali, Sirajganj during the period from January to December 2014. A total of 2646 clinical cases on cattle were recorded and analyzed. Diagnosis of each of the clinical cases was made on the basis of owner history, clinical examination and common laboratory techniques. The clinical cases were divided into three groups on the basis of treatment required viz. (1) Medicinal (2) Gynaeco-obstetrical and (3) Surgical cases. Among the three types of cases, medicinal cases constituted the highest percentage (79.33%) in comparison to gynaeco-obstetrical (11.60%) and surgical (9.07%) cases. Among the medicinal cases, the highest cases was recorded with parasitic diseases (55.97%), followed by infectious diseases (24.21%) and digestive disorders (10.34%). Other cases were general systemic states (3.91%), musculo-skeletal disorder (1.57%), skin condition (1.57%), metabolic diseases (1%), respiratory disorders (0.76%), sensory organ diseases (0.43%) and dog biting (0.24%). Among the gynaeco-obstetrical cases, repeat breeding (42.35%), anestrous (31.60%), orchitis (9.77%), posthitis (5.86%), dystocia (4.89%) and retained placenta (4.23%) were recorded as major gynaeco-obstetrical problems in cattle. Navel-ill (45%), myiasis (43.33%), abscess (6.25%) and fracture (3.34%) were recognized as the main disorders which required surgical interventions. Prevalence of diseases was high (39.38%) in summer season (March-June) followed by (34.73%) in winter (November-February) and lowest (25.89%) in rainy season (July-October). It may be concluded that a number of diseases with various percentages have been occurring in the Chauhali upazila and this report may help to develop control strategies against major diseases reported in this study.

INTRODUCTION

Livestock constitute an important part of the wealth of a country. It provides manure, meat and milk to the vast majority of the people. Cattle are a big portion of the livestock. There are about 23.4 million cattle in Bangladesh (Anon., 2014). About 20% of the population of Bangladesh earns their livelihood through work associated with raising cattle. Most of them are reared under smallholder traditional management system in rural areas. The management practices of animals and geo-climatic condition of Bangladesh are favorable for the occurrence of various diseases. Retrospective evaluation of clinical case records help to understand the predominant clinical problems and also their demographic and seasonal distribution in a particular area. Chauhali upazila of Siarajganj district in Bangladesh is surrounded by The Jamuna River and it is a natural calamity affected area which encourages many diseases in livestock. Although some reports on clinical case records from Bangladesh Agricultural University Veterinary Clinic (Rahman et al., 1972; Hossain et al., 1986; Das and Hashim, 1996; Samad, 2001; Samad et al., 2002), Haluaghat Upazila Veterinary Hospital, Mymensingh (Sarker et al., 1999) and Dairy Cooperatives in Pabna district (Pharo, 1987), Baghabari Milking zone of Bangladesh (Sarker et al., 2013), Ulipur Upazila Veterinary Hospital, Kurigram (Kabir et al., 2010), Khagrachari Sadar Veterinary Hospital, Khagrachari (Ali et al., 2011), Upazilla Veterinary Hospital, Mohammadpur, Magura (Karim et al., 2014), Chandanaish Upazila of Chittagong district, Bangladesh (Pallab et al., 2012) and Patuakhali Science and Technology University Veterinary Clinic (Rahman et al., 2012) are available but similar report on cattle are very limited in Chauhali upazila of Sirajganj district of Bangladesh. The objective was to determine the clinical prevalence of diseases and disorders in cattle at the Upazila Veterinary Hospital, Chauhali, Sirajganj.

MATERIALS AND METHODS

This clinical study was undertaken at the Upazila Veterinary Hospital (Officially named as upazilla livestock office), Chauhali, Sirajganj to determine the clinical prevalence of diseases and disorders in cattle during the one year study period from January to December, 2014.

General examination

Physical condition, behavior, posture, gait, superficial skin wound, prolapse of the uterus and vagina, salivation, nasal discharge, distension of the abdomen, locomotive disturbance etc were observed by visual examination of the patient.

Physical examination

Examination of different parts and system of the body of each of the sick animals were examined by using procedure of palpation, percussion, auscultation, needle puncture and walking of the animals.

Clinical examination

The temperature, pulse, and respiratory rate from each of these sick animals were recorded. Clinical examinations of all 2646 clinically sick cattle of different ages were conducted on the basis of diseases history, owner complaint, symptoms, to diagnose the following diseases and disorders. History of each case (present and past) was carefully taken which gave a guideline for examination of the animals. According the merit of the individual case, general clinical examination were conducted on the basis of disease history and owners complaint, symptoms and techniques such as microscopic examination, common laboratory techniques used by Rosenberger (1979) and Samad et al. (1988). These recorded clinical cases were primarily categorized into three major groups on the basis of treatment required. These groups were: (1) Medicinal cases (2) Gynaeco-obstetrical cases and (3) Surgical cases. The medicinal cases were categorized into major diagnostic groups that were considered sufficiently distinct so as to make clinical diagnosis accurate, such as (i) parasitic diseases (ii) infectious diseases (iii) digestive disorders (iv) general systemic states (v) musculo-skeletal disorders (vi) skin conditions (vii) metabolic and nutritional deficiency diseases (viii) respiratory disorders (ix) the sensory organ diseases and (x) other diseases. The study period was divided into three seasons on the basis of local climatic conditions viz. Summer (March to June), Rainy (July to October) and Winter (November to February). Data were organized in the Microsoft® Excel spreadsheet and percentages of disease conditions prevalent in different seasons were calculated.

RESULTS AND DISCUSSION

Of the 2646 recorded clinical cases of sick cattle, medicinal, gynaeco-obstetrical and surgical cases were 79.33%, 11.60% and 9.07% respectively (Table 1). This observation supports the earlier report of Rahman et al. (2012) and Karim et al. (2014). Rahman et al. (2012) recorded 84.1%, 4.7% and 11.20% and Karim et al. (2014) recorded 86.5%, 6.1% and 7.3% medicinal, gynaeco-obstetrical and surgical cases respectively. However, Samad (2001) recorded 90.76% medicinal, 5.46% gynaeco-obstetrical and 3.78% surgical cases in cattle. In addition to that, the highest number of cases were recorded in summer (39.38%), followed by winter (34.73%) and rainy (25.89%) seasons (Table 2, Figure 2), which is supported by Rahman et al. (2012) who reported the highest prevalence in summer followed by rainy and winter seasons.

Medicinal cases

Of the 2099 medicinal cases in cattle, the highest cases was recorded with parasitic diseases (55.97%), followed by infectious diseases (24.21%) and digestive disorders (10.34%). The least recorded cases were dog biting (Table 1). The findings support Rahman et al. (2012) who reported the highest 50.4% parasitic diseases, 14.2% digestive disorders, 14.8% systemic states and 5.5% respiratory disorders with some variations. However, infectious diseases are higher and systemic states are lower in this study because it is well known that the occurrence of disease varies with different geographical locations.

Parasitic diseases

The highest prevalence among different parasitic diseases was Gastro-intestinal worm infestation followed by Fascioliasis, Lice infestation and Coccidiosis (Table 1). The findings are supported by Rahman et al. (2012), Sarker et al. (2013), Kabir et al. (2010) and Ali et al. (2011) with a slight variation. The highest number of parasitic diseases may be due to grazing in the lowland area, irregular deworming, using inadequate dose of anthelmintics, favorable environment for the parasites etc. Again, the highest number of parasitic diseases in cattle was during winter followed by summer and rainy seasons (Table 2, Figure 1). However, Rahman et al. (2012) reported almost similar percentages of parasitic diseases during the three seasons in cattle.

Infectious diseases

The major infectious diseases in cattle were foot and mouth disease (FMD), ephimeral fever and dermatophilosis (Table 1). The findings are supported by Sarker et al. (2013), Kabir et al. (2010), Rahman et al. (2012) and Karim et al. (2014) with a slight variation. The highest frequency of FMD and other infectious diseases in the area is due to the fact that the area is situated beside a big river and is a flood affected area, introduction of diseased cattle from India etc. Again, the highest number of infectious diseases was recorded during summer season (Table 2, Figure 1) which is supported by Rahman et al. (2012).

Digestive disorders

Diarrhea (7.91%) and dysentery (2.43%) were found to be the major digestive disorders in cattle (Table 1). Rahman et al. (2012) and Sarker et al. (2013) also found the same result. However, Pallab et al. (2012) reported the highest 47.05% digestive disorders. Although the diarrheal and dysenteric cases were recorded in cattle in all the seasons of the year but highest percentage was recorded during summer followed by rainy and winter (Table 2, Figure 1). However, Rahman et al. (2012) recorded digestive disorders same during all seasons.

General systemic states

It was revealed that 2.48% and 1.43% cattle were affected with anorexia and acidosis (Tables 1). However, Sarker et al. (2013) reported 17.55% cattle with anorexia. In this study, most of the anorexic cases were counted under the specific diseases. The cases were recorded highest in rainy followed by summer and winter seasons (Table 2, Figure 1). However, Rahman et al. (2012) recorded the cases highest in winter followed by rainy and summer seasons.

Table 1. Clinical prevalence of diseases and disorders in cattle recorded at Upazilla Veterinary Hospital, Chauhali, Sirajganj

S/N	Diseases	Cattle (n=2646)	
		No. of affected cattle	Percentage (%)
	Parasitic diseases	**1175**	**55.97**
1	Gastro-intestinal Worm infestation	900	42.88
2	Fascioliasis	149	7.10
3	Lice infestation	70	3.33
4	Coccidiosis	28	1.33
5	Babesiosis	16	0.76
6	Hump sore	12	0.57
	Infectious diseases	**508**	**24.21**
7	FMD	204	9.72
8	Ephemeral Fever	111	5.29
9	Dermatophilosis	108	5.15
10	Mastitis	33	1.57
11	Papillomatosis	26	1.24
12	Black quarter	17	0.81
13	Tetanus	4	0.19
14	Foot rot	3	0.14
15	Actinobacillosis	2	0.10
	Digestive disorders	**217**	**10.34**
16	Diarrhoea	166	7.91
17	Dysentery	51	2.43
	General Systemic states	**82**	**3.91**
18	Anorexia	52	2.48
19	Acidosis	30	1.43
	Musculo-Skeletal disorder	**33**	**1.57**
20	Arthritis	33	1.57
	Skin condition	**33**	**1.57**
21	Urticaria	33	1.57
	Metabolic diseases	**21**	**1.00**
22	Agalactia	12	0.57
23	Milk fever	9	0.43
	Respiratory disorder	**16**	**0.76**
24	Pneumonia	16	0.76
	Sensory organ diseases	**9**	**0.43**
25	Corneal opacity	1	0.05
26	Otitis	8	0.38
	Other condition	**5**	**0.24**
27	Dog bite	5	0.24
Sub-total (Medicinal cases)		**2099**	**79.33**
1	Repeat breeding	130	42.35
2	Anestrous	97	31.60
3	Orchitis	30	9.77
4	Posthitis	18	5.86
5	Dystocia	15	4.89
6	Retention of placenta	13	4.23
7	Vaginitis	3	0.98
8	Abortion	1	0.33
Sub-total (Gyneco-obstetrical cases)		**307**	**11.60**
1	Navel-ill	108	45.00
2	Myiasis	104	43.33
3	Abscess	15	6.25
4	Fracture	8	3.34
5	Upward Patellar Fixation	3	1.25
6	Atresia ani	2	0.83
Sub-total (Surgical cases)		**240**	**9.07**
Overall		**2646**	**100**

Table 2. Season-wise Clinical prevalence of diseases and disorders in cattle recorded at Upazilla Veterinary Hospital, Chauhali, Sirajganj

S/N	Diseases	No. of affected cattle (%), Cattle (n=2646)			
		Summer	Rainy	Winter	Total
	Parasitic diseases	**429(36.51)**	**208(17.70)**	**538(45.79)**	**1175(55.97)**
1	Gastro-intestinal Worm infestation	356(39.56)	152(16.89)	392(43.56)	900(42.88)
2	Fascioliasis	27(18.12)	0	122(81.88)	149(7.10)
3	Lice infestation	28(40)	35(50)	7(10)	70(3.33)
4	Coccidiosis	9(32.14)	10(35.71)	9(32.14)	28(1.33)
5	Babesiosis	7(43.75)	5(31.25)	4(25)	16(0.76)
6	Hump sore	2(16.67)	6(50)	4(33.33)	12(0.57)
	Infectious diseases	**265(52.17)**	**177(34.84)**	**66(12.99)**	**508(24.21)**
7	FMD	119(58.33)	56(27.45)	29(14.22)	204(9.72)
8	Ephemeral Fever	43(38.74)	68(61.26)	0	111(5.29)
9	Dermatophilosis	78(72.22)	14(12.96)	16(14.81)	108(5.15)
10	Mastitis	10(30.30)	13(39.39)	10(30.30)	33(1.57)
11	Papillomatosis	4(15.38)	15(57.69)	7(26.92)	26(1.24)
12	Black quarter	9(52.94)	8(47.06)	0	17(0.81)
13	Tetanus	0	3(75)	1(25)	4(0.19)
14	Foot rot	0	0	3(100)	3(0.14)
15	Actinobacillosis	2(100)	0	0	2(0.10)
	Digestive disorders	**104(47.93)**	**63(29.03)**	**50(23.04)**	**217(10.34)**
16	Diarrhoea	87(52.41)	49(29.52)	30(18.07)	166(7.91)
17	Dysentery	17(33.33)	14(27.45)	20(39.22)	51(2.43)
	General Systemic states	**29(35.37)**	**34(41.46)**	**19(23.17)**	**82(3.91)**
18	Anorexia	24(46.15)	16(30.77)	12(23.08)	52(2.48)
19	Acidosis	5(16.67)	18(60)	7(23.33)	30(1.43)
	Musculo-Skeletal disorder	**9(27.27)**	**19(57.58)**	**5(15.15)**	**33(1.57)**
20	Arthritis	9(27.27)	19(57.58)	5(15.15)	33(1.57)
	Skin condition	**3(9.09)**	**17(51.52)**	**13(39.39)**	**33(1.57)**
21	Urticaria	3(9.09)	17(51.52)	13(39.39)	33(1.57)
	Metabolic diseases	**8(38.10)**	**7(33.33)**	**6(28.57)**	**21(1)**
22	Agalactia	5(41.67)	5(41.67)	2(16.67)	12(0.57)
23	Milk fever	3(33.33)	2(22.22)	4(44.44)	9(0.43)
	Respiratory disorder	**5(31.25)**	**3(18.75)**	**8(50)**	**16(0.76)**
24	Pneumonia	5(31.25)	3(18.75)	8(50)	16(0.76)
	Sensory organ diseases	**2(22.22)**	**3(33.33)**	**4(44.44)**	**9(0.43)**
25	Corneal opacity	0	1(100)	0	1(0.05)
26	Otitis	2(25)	2(25)	4(50)	8(0.38)
	Other condition	**2(40)**	**0**	**3(60)**	**5(0.24)**
27	Dog bite	2(40)	0	3(60)	5(0.24)
Sub-total (Medicinal cases)		**856(40.78)**	**531(25.30)**	**712(33.92)**	**2099(79.33)**
1	Repeat breeding	41(31.54)	18(13.85)	71(54.61)	130(42.35)
2	Anestrous	41(42.27)	28(28.87)	28(28.86)	97(31.60)
3	Orchitis	7(23.33)	16(53.33)	7(23.33)	30(9.77)
4	Posthitis	9(50)	5(27.78)	4(22.22)	18(5.86)
5	Dystocia	5(33.33)	6(40)	4(26.67)	15(4.89)
6	Retention of placenta	2(15.38)	5(38.46)	6(46.15)	13(4.23)
7	Vaginitis	0	1(33.33)	2(66.67)	3(0.98)
8	Abortion	0	1(100)	0	1(0.33)
Sub-total (Gyneco-obstetrical cases)		**105(34.20)**	**80(26.06)**	**122(39.74)**	**307(11.60)**
1	Navel-ill	27(25)	46(42.59)	35(32.41)	108(45)
2	Myiasis	47(45.19)	12(11.54)	45(43.27)	104(43.33)
3	Abscess	3(20)	12(80)	0	15(6.25)
4	Fracture	3(37.50)	2(25)	3(37.50)	8(3.34)
5	Upward Patellar Fixation	0	2(66.67)	1(33.33)	3(1.25)
6	Atresia ani	1(50)	0	1(50)	2(0.83)
Sub-total (Surgical cases)		**81(33.75)**	**74(30.83)**	**85(35.42)**	**240(9.07)**
Overall		**1042(39.38)**	**685(25.89)**	**919(34.73)**	**2646(100)**

Musculo-skeletal disorders

About 1.57% cattle were suffering from arthritis (Table 1). However, Sarker et al. (2013) reported 2.53% cases of arthritis in cattle. The highest number of arthritis in cattle was during rainy season (Table 2, Figure 1). However, Rahman et al. (2012) recorded the highest percentage of arthritis in cattle during summer season.

Skin conditions

Urticaria was recorded in 1.57% cattle (Table 1) and the highest cases were recorded during rainy seasons (Table 2, Figure 1). However, Ali et al. (2011) recorded 0.5% urticaria in cattle and the highest cases of urticaria in summer followed by winter and rainy season.

Metabolic diseases

Milk fever (0.43%) and agalactia (0.57%) was diagnosed in cattle under this group (Table 1). Ali et al. (2011) reported 0.59% milk fever and 2.09% agalactia. However, Pallab et al. (2012) reported 4.24% metabolic diseases. The frequency may be increased due to mineral deficiency or impairment of metabolism. The most of the cases were recorded in summer season (Table 2, Figure 1) which is supported by Rahman et al. (2012).

Respiratory disorders

Around 0.76% cases of pneumonia were recorded in this study (Table 1) which is supported by Samad (2001) and Karim et al. (2014) who reported 0.84% and 0.7% pneumonia in cattle, respectively. The highest percentage of pneumonia was recorded during winter season in cattle (Table 2, Figure 1) which is supported by Samad et al. (2002).

The sensory organ diseases

The cattle were affected with 0.38% otitis and 0.05% corneal opacity (Table 1). Sarker et al. (2013) reported 0.02% otitis and 0.97% eye disease. However, Samad et al. (2002) reported a high percentage (2.42%) of corneal opacity in cattle. The percentage of eye and ear diseases was the highest during winter season (Table 2, Figure 1). However, Samad et al. (2002) reported the highest percentage of eye diseases during rainy season.

Dog biting

Dog bite was recorded in 5 cattle (0.24%). The cases were recorded as 60% in winter season and 40% in summer season (Table 2, Figure 1). Ali et al. (2011) found one case of dog biting only in summer season during a five year study.

Gynaeco-obstetrical cases

Repeat breeding

Repeat breeders are those cows that fail to conceive after three or more regularly spaced services in the absence of detectable abnormalities of the internal genitalia (Samad, 2000). The highest gyneco-obstetrical cases, repeat breeder was recorded in 42.35% cattle (Table 1). Karim et al. (2014) also reported the same result. Public unconsciousness, no service in the remote area, quack treatment etc. are some of the major causes of repeat breeder cattle. The highest number of repeat breeding in cattle was found during winter season (Table 2, Figure 2) which is supported by Rahman et al. (2012).

Anestrous

Anestrous was recorded in 31.60% cows (Table 1). Rahman et al. (2012) reported 59.50% anestrus cases in cattle. Vitamin A deficiency, cystic ovaries, atrophied ovaries, other ovarian disease are mainly responsible for anestrous in cattle. The highest number of cases in cattle was recorded during summer seasons (Table 2, Figure 2), However, Rahman et al. (2012) reported the highest in winter season.

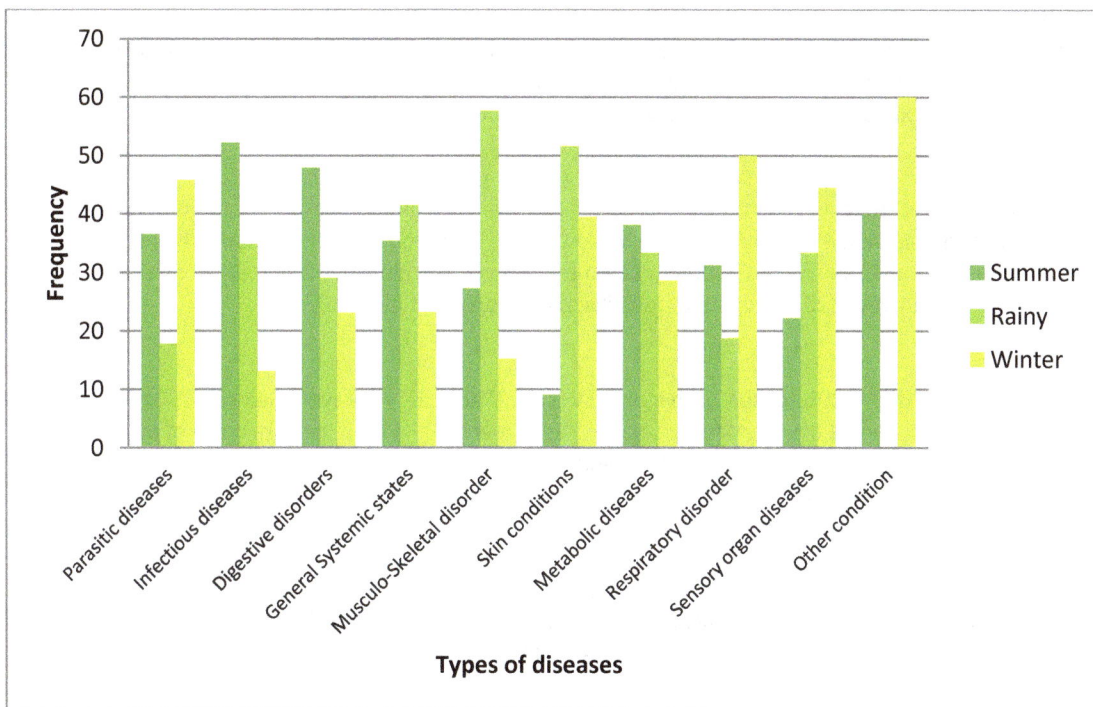

Figure 1. Season-wise distribution of various medicinal cases in cattle

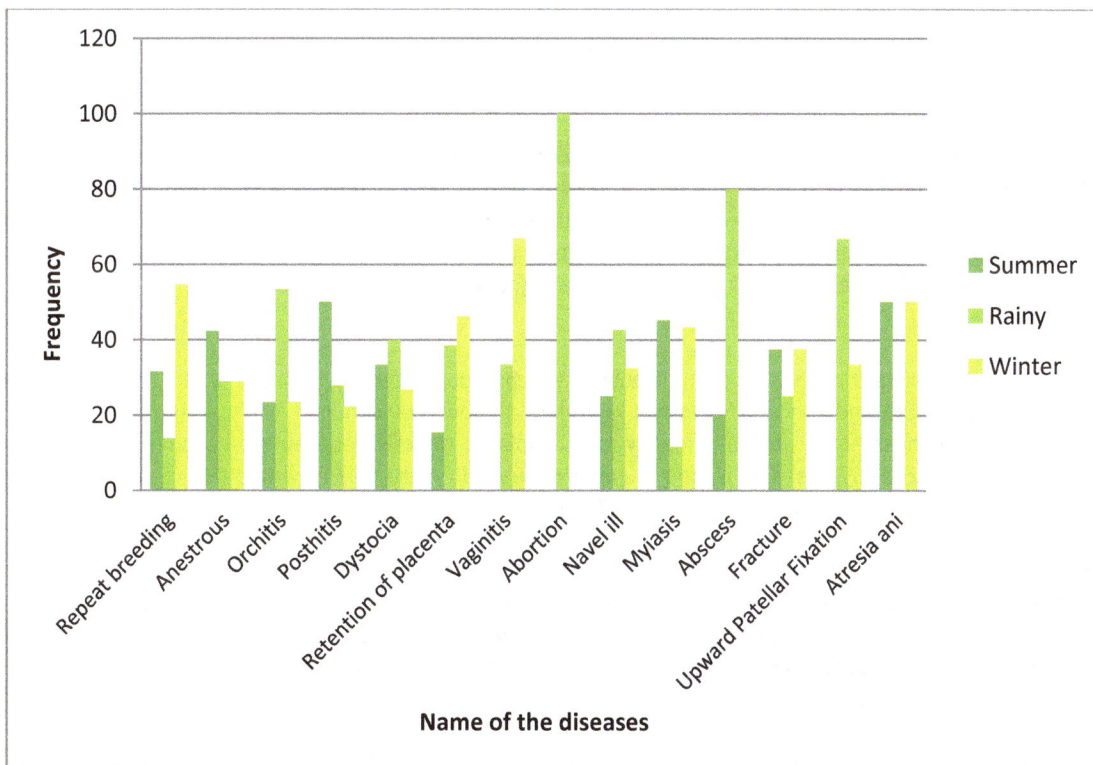

Figure 2. Season-wise distribution of various gyneco-obstetrical and surgical cases in cattle

Orchitis

This study recorded 9.77% cases of orchitis in cattle (Table 1). The highest number of orchitis in cattle was during rainy, followed by summer and winter seasons (Table 2, Figure 2). However, the recorded cases in cattle could not be compared due to lack of similar inland reports.

Posthitis

Posthitis was recorded in 5.86% cattle (Table 1). However, Rahman et al. (2012) reported the cases 0.70% in cattle. In this study, most of the cattle were bulls so frequency of the disease is higher. The highest number of cases in cattle was recorded during summer, followed by rainy and winter seasons (Table 2, Figure 2). However, Rahman et al. (2012) reported posthitis in all seasons of the year.

Dystocia

This study recorded 4.89% cases of dystocia in cows (Table 1). However, Samad (2001) and Ali et al. (2011) reported 0.02% and 1.8% dystocia cases in cows respectively. The highest number of dystocia in cattle was during rainy, followed by summer and winter seasons (Table 2, Figure 2). However, Rahman et al. (2012) recorded dystocia only in summer season.

Retained placenta

This disorder was recorded only in 4.23% cows (Table 1). Rahman et al. (2012) reported 8.1% cases of retained placenta in cows. In winter seasons the cases were mostly found (Table 2, Fig. 2). Rahman et al. (2012) recorded the most of the cases of retained placenta in rainy season.

Vaginitis and abortion

These disorders were recorded only in 0.98% and 0.33% cows respectively (Table 1). Ali et al. (2011) reported 0.90% abortion in cows. Vaginitis is recorded the highest in winter followed by rainy seasons. The abortion cases were recorded only in summer season in cattle (Table 2, Figure 2).

Surgical cases

Navel-ill

The highest surgical cases, navel-ill were recorded in 45% cattle (Table 1). Sarker et al. (2013) and Sarker et al. (2014) also found navel ill was the highest among the surgical disorders. Unhygienic maternity pen and calving pen, lack of colostrums intake, no antiseptic use on the naval cord, unconsciousness may be responsible for the high frequency of navel-ill. The highest cases were recorded during rainy season (Table 2, Figure 2). However, Rahman et al. (2012) recorded the highest cases during winter season.

Myiasis

Myiasis was recorded in 43.33% cattle (Table 1) which is supported by Rahman et al. (2012). However, Nooruddin et al. (1986) and Das and Hashim, (1996) found a low prevalence of 1.07% and 2.20% myiasis in cattle. Unconsciousness of the people is mainly responsible for myiasis because the abscess is not properly treated which produces myiasis. The highest number of cases in cattle was recorded during summer season (Table 2, Figure 2) which is supported by Samad (2001) and Rahman et al. (2012).

Abscess

Abscess was recorded in 6.25% cattle (Table 1). This observation supports the report of Sarker et al. (2013). However, Rahman et al. (2012) reported 1.1% cattle affected with abscess. The highest cases were recorded during rainy season and rest in summer season (Table 2, Figure 2). Rahman et al. (2012) recorded abscess only in summer season.

Fracture

Fracture was recorded in 3.34% cattle (Table 1). The number of fracture cases in cattle was recorded as same during summer and winter seasons followed by rainy season (Table 2, Figure 2). However, the recorded cases in cattle could not be compared due to lack of similar inland reports.

Upward Patellar Fixation

Upward Patellar Fixation was recorded in 1.25% cattle (Table 1) and the cases were recorded during rainy and winter seasons (Table 2, Figure 2). Rahman et al. (2012) supports the finding who reported 2.2% cattle affected with upward patellar fixation but only in summer season.

Atresia ani

This study recorded 0.83% cases of atresia ani in calves (Table 1). The cases were recorded as half in summer and half in winter seasons (Table 2, Figure 2). However, the recorded cases in cattle could not be compared due to lack of similar inland reports.

CONCLUSIONS

Occurrence of diseases was recorded during clinical examination of sick cattle at Upazila Veterinary Hospital, Chauhali, Sirajganj, Bangladesh. It was observed from the study that the cattle were most susceptible to parasitic infestation. Parasitic infestation causes heavy economic losses in every year. So, regular anthelmintics treatment should be given to control the parasitic diseases. Proper planning and program should be undertaken to prevent and control diseases and disorders of cattle in the study area.

AUTHORS' CONTRIBUTIONS

MG Morshed and SHMF Siddiki designed the experiments, collected the data from the hospital, MS Parvin and SHMF Siddiki analysed the data. MS Parvin, L Naher and SHMF Siddiki wrote the manuscript. All authors participated in experimental design and read and approved the final manuscript.

REFERENCES

1. Ali MH, MKJ Bhuiyan and MM Alam, 2011. Retrospective epidemiologic study of diseases in ruminants in Khagrachari hill tract district of Bangladesh. Bangladesh Journal of Veterinary Medicine, 9: 145-153.
2. Das BC and MA Hashim, 1996. Studies on surgical affections in calves. Bangladesh Veterinary Journal, 30: 53-57.
3. Debnath NC, BK Sil, SA Selim, MAM Prodhan and MMR Howlader, 1990. A retrospective study of calf mortality and morbidity on smallholder traditional farms in Bangladesh. Preventive Veterinary Medicine, 9: 1-7.
4. Anon., 2014. http://en. banglapedia. org/ index. php? title= Cattle.
5. Hoque MS and MA Samad, 1996. Prevalence of clinical diseases in dairy cross-bred cows and calves in the urban areas in Dhaka. Bangladesh Veterinary Journal, 30: 118-129.
6. Hossain MA, M Shaidullah and MA Ali, 1986. A report on surgical diseases and reproductive disorders recorded at the Veterinary Hospital of Bangladesh Agricultural University, Mymensingh. Bangladesh Veterinary Journal, 20: 1-5.
7. Kabir MH, MA Reza, KMA Razi, MM Parvez, MAS Bag and SU Mahfuz, 2010. A report on clinical prevalence of diseases and disorders in cattle and goat at the Upazilla Veterinary Hospital, Ulipur, Kurigram. International Journal of Biological Research, 2: 17-23.
8. Karim MR, MS Parvin, MZ Hossain, MT Islam and MT Hussan, 2014. A report on clinical prevalence of diseases and disorders in cattle and goats at the Upazilla Veterinary Hospital, Mohammadpur, Magura. Bangladesh Journal of Veterinary Medicine, 12: 47-53.
9. Nooruddin M and AS Dey, 1990. Further study on the prevalence of skin diseases in domestic ruminants of Bangladesh. Bangladesh Veterinarian, 7: 75-81.
10. Nooruddin M, AJ Sarker, MA Baki, MR Ali, AS Dey and MF Hoque, 1986. Prevalence of diseases of external organs of cattle. Bangladesh Veterinary Journal, 20: 11-16.
11. Pallab MS, SM Ullah, MM Uddin and OF Miazi, 2012. A cross sectional study of several diseases in cattle at Chandanaish Upazilla of Chittagong district. Scientific Journal of Veterinary Advances, 1: 28-32.

12. Pharo HJ, 1987. Analysis of clinical case records from dairy co-operatives in Bangladesh. Tropical Animal Health and Production, 19: 136-142.
13. Rahman A, JU Ahmed and MA Haque, 1975. Analysis of lameness of cattle admitted to the Veterinary hospital of Bangladesh Agricultural University. Bangladesh Veterinary Journal, 9: 21-24.
14. Rahman MA, KM Ali and A Rahman, 1972. Incidence of diseases of cattle in Mymensingh. Bangladesh Veterinary Journal, 6: 25-30.
15. Rahman MA, MA Islam, MA Rahman, AK Talukder, MS Parvin, and MT Islam, 2012. Clinical diseases of ruminants recorded at the Patuakhali Science and Technology University Veterinary Clinic. Bangladesh Journal of Veterinary Medicine, 10: 63-73.
16. Rahman MM, M Ali and A Hashem, 1999. Livestock disease problems in a selected area of Sherpur district. Bangladesh Journal of Training and Development, 12: 205-210.
17. Rosenberger G, 1979. *Clinical Examination of Cattle.* 2nd edn, Verlag Poul Parey, Germany.
18. Samad MA, 2000. *Veterinary Practitioner's Guide.* 1st Publication, LEP Publication No. 07, BAU Campus, Mymensingh.
19. Samad MA, 2001. Observations of clinical diseases in ruminants at the Bangladesh Agricultural University Veterinary Clinic. Bangladesh Veterinary Journal, 35: 93-120.
20. Samad MA, ASM Bari and SA Bashar, 1988. Gross and histopathological studies on bovine babesiosis in Bangladesh. Indian Journal of Animal Science, 58: 926-928.
21. Samad MA, MA Islam and A Hossain, 2002. Patterns of occurrence of calf diseases in the district of Mymensingh in Bangladesh. Bangladesh Veterinary Journal, 36: 01-05.
22. Sarker MAS, M Aktaruzzaman, AKMA Rahman and MS Rahman, 2013. Retrospective study of clinical diseases and disorders of cattle in Sirajganj district in Bangladesh. Bangladesh Journal of Veterinary Medicine, 11: 137-144.
23. Sarker MAS, MA Hashim, MB Rahman and H Begum, 1999. Studies on bovine lymphadenitis syndrome. Bangladesh Veterinarian, 10: 6-8.
24. Sarker NU, K Samaddar, MM Haq and MM Rahman, 2014. Surgical affections of cattle in the milk-shed areas of Bangladesh. The Bangladesh Veterinarian, 31: 38 – 45.

TRANSBOUNDARY DISEASES OF ANIMALS: CONCERNS AND MANAGEMENT STRATEGIES

M. Ariful Islam*

Department of Medicine, Faculty of Veterinary Science, Bangladesh Agricultural University, Mymensingh-2202, Bangladesh

*Corresponding author: M. Ariful Islam; E-mail: maislam77@bau.edu.bd

---

ARTICLE INFO	ABSTRACT

Transboundary animal diseases (TADs) are greatly contagious epidemic diseases that can spread very rapidly, irrespective of national borders. They cause high rates of death and disease in animals, thereby having serious socio-economic and sometimes public health consequences while constituting a steady threat to the livelihoods of livestock farmers. With the development of technology, livestock production has gained an integral position in the national economy, socioeconomic development, poverty alleviation and nutrition supply for human. Livestock farming is one of the important sources of livelihood to rural peoples in Bangladesh. A healthy livestock is pleasure of any country including Bangladesh. However, rapid trend of globalization has brought upon challenges in maintaining healthy herds of livestock. The emerging infections of foreign origin could spread across national geographical borders and cause devastation in livestock population. As a result, there will be an emergence and spread of new disease in the region which was once free from the disease. Regional and international approaches have to be followed, and the FAO and OIE Global framework-TADs initiative provides the suitable concepts and objectives as well as an organizational framework to link international and regional organizations at the service of their transboundary diseases.

Key words

Livestock health,
Global challenge,
Transboundary
animal disease
(TAD),
Global trade,
Management
strategy

economic impact of TADs and emerging animal diseases. In this paper, we have summarized the main diseases of livestock that are transboundary in nature, and sum up the challenges and necessary management strategies in controlling the

INTRODUCTION

Transboundary animal diseases (TADs) pose a serious risk to the world animal agriculture and food security and jeopardize international trade. The world has been facing devastating economic losses from major outbreaks of TADs. It is unimaginable to have a human society without a healthy population of livestock. Livestock not only provide food security but also improve the quality of human life and make a significant contribution to national economy. Several thousands of small and marginal farmers in the country depend solely on agricultural farming and livestock husbandry. The existence of infectious diseases affecting farm animals has been historically recorded for over hundreds of years. However, factors associated with modernization of human societies such as changes in agro-ecological conditions and global marketing, have led to increased incidences of animal diseases. This is mainly due to spread of disease causing pathogens across borders. With increasing movement of human population, livestock and livestock products, fish and fish products, and plants and plant products within and across countries, together with climate changes, threat from transboundary diseases is intensifying. Transboundary diseases are highly contagious and have the potential for rapid spread, irrespective of national borders, causing serious socioeconomic consequences (Otte et al., 2004). Traditionally, trade, traffic and travel have been instruments of disease spread. Now, changing climate across the globe is adding to the misery. Climate change is creating new ecological platform for the entry and establishment of pests and diseases from one geographical region to another (FAO, 2008). Several new transboundary diseases emerge, and old diseases reemerge, exhibiting increased chances for unexpected spread to new regions, often over great distances.

Transboundary livestock diseases such as Foot-and-mouth disease (FMD) have a direct economic impact by reducing agricultural and animal production (FAO/OIE, 2004; Domenech et al., 2006). Apart from causing suffering and mortality in susceptible population, the diseases adversely affect food safety, rural livelihoods, human health and international trade. Therefore, it is necessary to effectively manage the transboundary diseases. In developing countries, control of these diseases is a key pathway for poverty alleviation. It is advisable to have an effective quarantine system in place to prevent entry and establishment of trans-boundary diseases. As a second line of defense, a country must also have in place a suitable contingency plans to respond quickly to high threat diseases (Basagoudanavar and Hosamani, 2013). This could be achieved by timely application of scientific technology for rapid response. A disease outbreak in the neighboring country should always be taken as an immediate threat. Thus, it's a big challenge in managing and controlling TADs without collective and collaborative action between neighboring countries.

TRANS-BOUNDARY ANIMAL DISEASES (TADs)

Transboundary animal diseases are permanent global threat for livestock farmers. TADs are defined as:

"Those that are of significant economic trade and/or food security importance for a considerable number of countries; which can easily spread to other countries and reach epidemic proportions; and where control management, including exclusion requires cooperation between several countries".

Within theses definition, there are many diseases that cause damage and destruction to farmers' property, may threaten food security, injure rural economies, and potentially disrupt trade relations.

The common ways of introduction of animal diseases to a new geographical location are through entry of live diseased animals and contaminated animal products. Other introductions result from the importation of contaminated biological products such as vaccines or germplasm or via entry of infected people (in case of zoonotic diseases). Even migration of animals and birds, or natural spreading by insect vectors or wind currents, could also spread diseases across geographical borders. All animal diseases have the potential to adversely affect human populations by reducing the quantity and quality of food, other livestock products (hides, skins, fibers) and animal power (traction, transport) that can be obtained from a given quantity of resources and by reducing people's assets. Of these, transboundary animal diseases tend to have the most serious consequences. TADs may be defined as those epidemic diseases which are highly contagious or transmissible and have the potential for very rapid spread, irrespective of national borders, causing serious socio-economic and possibly public health consequences. These diseases which cause a high morbidity and mortality in susceptible animal populations, constitute a constant threat to the livelihood of livestock farmers.

Furthermore, their potential consequences are of such a magnitude that their occurrence may also have a significant detrimental effect on national economies. TADs have the potential to:

- Threaten food security through serious loss of animal protein and/or loss of draught animal power for cropping;

- Increase poverty levels particularly in poor communities that have a high incidence dependence on livestock farming for sustenance;

- Cause major production losses for livetsock products such as meat; milk and other dairy products; wooland other fibers and skins and hides, thereby reducing farm incomes. They may also restrict opportunities for upgrading the production potential of local livestock industries by making it difficult to utilise exotic high producing breeds which tend to be very susceptible to the transboundary disease;

- Add significantly to the cost of livestock production through the necessity to apply costly disease control measures;

- Seriously disrupt or inhibit trade in livestock and livestock products either within a country or internationally. Their occurrence may thereby cause major losses in national export income in significant livestock-producing countries;

- Cause public health consequences in the case of those transboundary animal diseases which can be transmitted to humans (i.e. zoonoses);

- Cause environmental consequences through die-offs in wildlife populations in some cases, and;

- Cause pain and suffering for affected animals.

The most common TADs (Otte et al., 2004; FAO/OIE, 2004) under this category are provided in Table 1.

Table 1: Major Trans-boundary Animal Diseases.

Disease	Animals affected	Regions with major incidence
Foot-and-mouth disease (FMD)	Cattle, buffalo, sheep, goats and pigs	Parts of Africa, Middle East and Asia
Peste des petits ruminants (PPR)	Sheep and goats	Africa, Middle East and Asia
Classical swine fever (CSF)	Pigs	South and South-East Asia
African swine fever (ASF)	Pigs	Sub-Saharan Africa, West Africa, parts of Europe and Latin America
Blue tongue (BT)	Sheep, cattle	Australia, USA, Africa, Middle East, Asia and Europe
Rift Valley Fever (RVF)	Sheep, cattle and goats	Africa
Contagious bovine pleuropneumonia (CBPP)	Cattle	Eastern, Southern and West Africa, parts of Asia
Lumpy skin disease (LSD)	Cattle	Africa
Sheep and goat pox	Sheep and goats	South Asia, China, Middle East, Africa
Bovine spongiform encephalopathy (BSE)	Cattle	UK and other parts of Europe
Venezuelan Equine Encephalomyelitis	Equines	Central American and South American countries
Newcastle disease (ND)	Poultry	Asia and Africa
Highly pathogenic avian influenza (HPAI)	Poultry	Asia, Europe and Africa
Hendra virus (HeV) infection	Horses	Australia
Nipah virus (NiV) infection	Pigs	Malaysia and Singapore

CHALLENGES IN DEALING WITH TADs

TADs are permanent threat for livestock keepers. They have major economic implications: both through the private and public costs of the outbreak, and through the cost of the measures taken at individual, collective and international levels in order to prevent or control infection and disease outbreaks. Several challenges confront the strategies to combat TADs (FAO, 2008; Hitchcock et al., 2007). The major ones are presented below:

- Requirement of novel systems having capacity of real-time surveillance of emerging diseases. For this, need driven research and service oriented scientific technology are a necessary at regional levels. Research emphasis has to be on specific detection and identification of the infectious agents;
- Need for epidemiological methods to assess the dynamics of infections in the self and neighboring countries/regions. These methods should be of real-time utility;
- Need for research and development of disease diagnostic reagents those do not need refrigeration (cold chain). More importantly, they should be readily available as well as affordable, for use in pen-side test format;
- There are many diseases for which there is inadequate supply of vaccines or there are no vaccines available. Insufficient or lack of vaccine hampers the disease control programs. Need to build up vaccine banks for stockpiling the important vaccines to implement timely vaccination;
- Required availability of cost-effective intervention or disease control strategies. Even if a technology is available, it has to be cheaper to adopt at the point of use;
- Need for ensuring public awareness of epidemic animal diseases. Many farmers are unaware of the emerging diseases. As such, unless reported to concerned regional authority, an emerging disease may go unnoticed;
- Shortage of government and private funding for research on emerging animal disease problems. Government as well as industries dealing with animal health should take initiative and appropriate sponsorship in this regard, and;
- Inadequate regulatory standards for safe international trade of livestock and livestock products. Otherwise, there would be a compromised situation in disease control strategies.

Management strategies of TADs

FAO animal disease emergency response mechanisms (FAO, 2008)

The FAO Emergency Prevention System (EMPRES) Animal Health develops strategies for intervention and improved management. It works to monitor and give early warning and ultimately to prevent animal diseases.

The Emergency Centre for Transboundary Animal Diseases (ECTAD) is FAO's corporate centre for the planning and delivery of veterinary assistance to FAO member countries responding to the threat of transboundary animal health crises.

The Crisis Management Centre-Animal Health is FAO's rapid response unit to animal disease emergencies. For example, highly Pathogenic Avian Influenza – bird flu. Since 2004 FAO has been at the forefront of the fight against highly pathogenic avian influenza (HPAI) – bird flu – in over 95 countries. FAO has mobilized over US$ 445 million to combat influenza and emerging disease threats through prevention, surveillance, and control.

Various strategies need to be implemented to prevent and control trans-boundary diseases regionally and internationally. These include:

- Preventing incidence of trans-boundary diseases and disease transmitting vectors. Minimizing the movement of animals across the borders is essential. Also, prompt practice of quarantine 2013 protocol would reduce many transboundary diseases. Geographic information system (GIS) and remote sensing could be utilized as early warning systems and in the surveillance and control of infectious diseases (Martin et al., 2007);

- Reducing man-made disasters that have adverse implications on climate. Global warming and climate change either due to natural or anthropogenic influences are likely to predispose the animal population to newer infections (FAO, 2008). Therefore collective efforts are needed to minimize adverse climatic changes;
- Interrupting the human-livestock wildlife transmission of infections. Diseases at the wildlife–livestock interface must become the focus for surveillance of emerging infectious diseases (Siembieda et al., 2011). Breaking the cycle of disease transmission would help control the spread of infections;
- Establishing regional biosecurity arrangement with capacity for early disease warning system for surveillance, monitoring and diagnosis of emerging disease threats (Domenech et al., 2006);
- Undertaking animal breeding strategies to create disease resistant gene pools. Enhancing host genetic resistance to disease by selective breeding of resistant animals is a smart strategy to improve natural immunity of animals to counter invading infections (Gibson et al., 2005);
- Strengthening government policies to enhance agricultural/animal research and training, and technology development (Rweyemamu et al., 2006). More funds need to be allocated for this purpose to build goal oriented research programs in combating TADs;
- Ensuring appropriate preparedness and response capacity to any emerging disease. Keeping in view that emerging infectious diseases are a constant threat, it is necessary to have early disease detection capacity and then implement a timely response (Hitchcock et al., 2007);
- Intensification of international cooperation in preventing spread of TADs. As TADs are a concern globally, cumulative effort is needed at international level to minimize the spread of infectious diseases across the borders (Domenech et al., 2006; Hitchcock et al., 2007).

SUMMARY

Effective TAD management and control is possible, and has been achieved by several countries in the World including two Asian countries (Japan and Korea), to do so it is highly dependent on national governance effective veterinary services and political support. Improved veterinary services and strong regional networks and have the choice to improve their own leadership and management. It is within the control of a veterinary service to develop a TAD control strategy and operational plans. Furthermore, these might be added to increase the capacity indicators such as: a strong relationship with commercial operators, NGOs and civil society; mechanisms for overcoming the constraints to TAD control posed by decentralisation; innovative financing mechanisms backed up by accountability and transparency; continued development of professionals in public and private service; and strong regional collective efforts. With rapidly increasing globalization, an associated risk of movement of trans-boundary diseases is emerging. Trans-boundary animal diseases represent a serious threat. They reduce production and productivity, disrupt local and national economies, and also threaten human health. This imposes extensive challenges for agricultural scientists on the critically important need to improve technologies in animal production and health in order to make sure the food security, poverty alleviation and to aid economic growth. Considering that livestock rearing constitutes a significant share in the national economy of a developing country like ours, it is imperative to take up disease control initiatives. Measures are required to safeguard the livestock industry from epidemics of infectious diseases and to uphold safe international trade of livestock and their products. In this regard, it is essential to develop scientific and risk-based standards that facilitate the international trade in animal commodities. There is also a need for information on regional movements of animals, birds and poultry products, in order to plan regional strategies of TADs control.

REFERENCES

1. Basagoudanavar SH and M Hosamani, 2013. Trans-boundary Diseases of Animals: Mounting Concerns, VetScan, Vol. 7, No. 2.
2. Domenech J, Lubroth J, Eddi C, Martin V, Roger F, 2006. Regional and international approaches on prevention and control of animal trans-boundary and emerging diseases.
3. FAO, 2008. Expert meeting on climate related trans-boundary pests and diseases including relevant aquatic species. FAO headquarters, 25-27 February 2008. Rome, Italy.

4. FAO/OIE, 2004. Joint FAO/OIE initiative. The global framework for the progressive control of trans-boundary animal diseases (GF-TADs).

5. Gibson JP and Bishop SC, 2005. Use of molecular markers to enhance resistance of livestock to disease: A global approach. OIE Sci. Technolgy Review, 24: 343-353.

6. Hitchcock P, Chamberlain A, Van Wagoner M, Inglesby TV and O'Toole T, 2007. Challenges to global surveillance and response to infectious disease outbreaks of international importance. Biosecurity and Bioterrorism, 5: 206-227.

7. Martin V, De Simone L, Lubroth J, 2007. Geographic information systems applied to the international surveillance and control of trans-boundary animal diseases, a focus on highly pathogenic avian influenza. Veterinaria Italiana, 43: 437-450.

8. Otte MJ, Nugent and McLeod A, 2004. Trans-boundary animal diseases: Assessment of socio-economic impacts and institutional responses. Livestock policy discussion paper No. 9. FAO, Rome, Italy.

9. Rweyemamu MM, Musiime J, Thomson G, Pfeiffer D and Peeler E, 2006. Future control strategies for infectious animal diseases-Case study of the UK and sub-Saharan Africa. In: UK Government's foresight project, Infectious diseases: preparing for the future, pp 1-24.

10. Siembieda JL, Kock RA, McCracken TA and Newman SH, 2011. The role of wildlife in trans-boundary animal diseases. Animal Hlth. Research Reviews, 12: 95-111.

7

EFFECT OF INDIGENOUS HERBAL PREPARATIONS ON COCCIDIOSIS OF POULTRY

ABM Jalal Uddin[1*], Maksudur Rashid[1], Md. Anwarul Khan[2], Md. Abdul Awal[2], Md. Abdus Sobhan[2] and AWM Shamsul Islam[3]

[1]Department of Livestock Services, Bangladesh; [2]Department of Pharmacology and [3]Department of Parasitology, Faculty of Veterinary Science, Bangladesh Agricultural University, Mymensingh-2202, Bangladesh

*Corresponding author: ABM Jalal Uddin; E-mail: abulbasar47@gmail.com

ARTICLE INFO

Key words

Coccidiosis

Phytotherapy

Chicken

ABSTRACT

Coccidiosis is a protozoan disease in chickens caused by *Eimeria spp* with great economic significance. The disease can be controlled by using modern anticoccidial drugs. However large scale and long term use of anticoccidial drugs has led to the resistance. Phytotherapy can be used successfully as an alternative coccidiosis control strategy. Ninety six chicks experimentally infected with coccidiosis were used in the present study. The chicks were divided into 24 groups and were treated with the crude watery extract (10%) of mango, pineapple, guava, chutra leaves and thankuni at the dose rate of 1ml, 5ml and 10ml per Kg body weight as well as one group was treated with Embazine® at recommended dose. The effects of these drugs were evaluated by oocyst count from faecal sample and the mortality rate. The result shows that the 10ml/Kg dose worked better than other dose. In this dose oocyst count significantly decreased (P< 0.01) at the day of 4 (mango), 2 (pineapple), 12 (guava), 2 (chutra) and 4 (thankuni). No oocyst found in feces at the day 8 (chutra and thankuni), 12 (mango and pineapple) whereas on day 12 it shows oocyst in feces in case guava. It shows that mortality encounter 75% in mango and guava group, 50% in pineapple and thankuni group, wheras no mortality recorded in chutra group. After completion of experiment it was noted that chutra leaves was most effective at the dose rate of 10 ml/kg body weight and effective near about Embazine.

INTRODUCTION

Coccidiosis is a common infectious disease in poultry, causing major economic losses. The protozoan parasite of the genus *Eimeriaspp* in the cecum and intestinal tract of poultry and produces tissue damage, resulting in reduced growth and increased susceptibility to pathogens (McDougald, 2003) such as *Clostridium perfringens*, leading to necrotic enteritis (Helmboldt and Bryant, 1971; Maxy and Page, 1977; Shane *et al.,* 1985). Coccidiosis is considered as one of the serious problems for poultry development in our country. Srinivasan (1959) reported 90- 100% mortality in chicken to be associated with coccidiosis in India. Because of favourable climatic condition this disease is prevalent seriously in Bangladesh also. Mortality in young birds which varies from 25-90% days of infection is predominant factor of economic loss (samad 1988). If we can establish the traditional system of medicine it will be highly benificial for the farmers and for the overall improvement of the livestock.

The world health organization (WHO) has recognized the necessity for investing and mobilizing the ancient medicinal practice for meeting the primary health needs of the people, the traditioal systems will provide basic health services to the rural population of the developing country like ours. No systemic assessment of the impact of traditional system on livestock health has yet been attempted, nor does it have an organized apporach at present. Due to vast usage of sulphanilamide, amprolium or synthetic chemical compounds for the treatment of coccidiosis in poultry results in emergence of drug resistant anticocccidial herbs preparations are required. Several herbs possess anti coccidial effects such as *urticadicia, mangoferousindica, hydrocotylasiatica* etc.So this study of the indigenous system of herbal preparation (mango, pineapples, guava, chutra and thankuni) has been undertaken to determine the anticoccidial effects in chickens.

MATERIALS AND METHODS

The experiment was performed in the Department of Pharmacology, Bangladesh Agricultural University, Mymensingh, Bangladesh.Ninety six chicks of white leg horn breed infected with coccidiosis and weighing between 250-350 gms were used for this study during March to June 1989. All the chicks were 4 weeks of age. At first ten chicks suffering from coccidiosis were collected from the private farm. From these diseased chicks, coccidiosis was produced in rest of the chicks. All the infected coccidiosis chicks were kept in good housing condition free from any contamination and they were maintained on a balance ration consisting of 3200 k cal/kg feed and 20% protein and water *ad lib.*

The procedure employed in the present study consisted of 3 steps. Viz: A. Collection and preparation of crude watery extract (10%) of medicinal plants; B. Studies on the efficacy of different crude watery extracts of medicinal plants against coccidiosis in chicks; C. Studies on the comparative efficacy of crude watery extract of selected medicinal plants with that of modern patent drug Embazine against cocccidiosis in chicks.

A. Collection and preparation of crude watery extract (10%) of medicinal plants.

The leaves of selected plants such as mango, pineapple, guava, chutra and the whole of thankuni were seleted in fresh condition. The collected growing mango, pineapple, guava, chutra and the whole of thankuni were washed thoroughly with water. Then these were cut into small pieces and dried in the sun for 7 days. These dried sample were taken into equal parts by measuring with balance and pulverized with the help of a grinding machine. Finally the ground power was preserved in an air tight bottle. 200 gms powders of fresh leaves of each plants was prepared.From the ground powder of the sample 10 gms from each sample was taken in 100ml of distilled water in a clean beaker and mixed thoroughly by a stirrer. Then the beaker was placed on a heater and the content of the beaker was boiled for 15-20 minutes. After replace from the heater the contents were filtered by passing through fine muslin. The watery extracts was then placed in a sterilized air tight bottle and labeled for further study.

For administration to the chick, fresh crude watery extracts were prepared daily. Each kind of samples was kept in separate glass bottle with labeling.

B. Studies on the efficacy of different crude watery extracts of medicinal plants against coccidiosis in chicks.

Eighty infected (coccidiosis) chicks were used for this study. The disease condition of coccidiosis of the chicks were confirmed by observing visiable signs and symptoms and also by examining the presence of oocyst in the faeces. All the infected chicks were randomly divided into 20 groups each consisting of 4 chicks. All the chicks were housed group by group in 20 different cages. The cages were numbered as I to XX. Group I chicks did not receive any treatment and was kept as control. Group II, III, and IV chicks were treated with crude watery extract of mango leaves at the dose rate of 1ml, 5ml and 10 ml per kg body weight respectively.

Groups VI,VII and VIII chicks received crude watery extracts of pine apple leaves at the dose rate of 1ml,5ml and 10ml per kg body weight respectively. Group V chicks were kept as control without giving any treatment.Similarly group IX chicks did not receive any treatment and was kept as control. Group X, XI, XII, chicks received crude watery extracts of guava leaves at the dose rate of 1ml, 5ml and 10 ml per kg body weight respectively.Likewise groups XIV, XV and XVI chicks were given crude watery extract of chutra leaves at the dose rate of 1ml, 5ml and 10 ml per kg body weight respectively. Here chicks of group XIII was kept as control without providing any treatment. In the last group the chicks of group XVII was kept as control which received no treatment. Crude watery extracts of thankuni leaves at the dose rate of 1ml, 5ml and 10 ml per kg body weight were administered to groups XVIII, XIX and XX chicks, respectively.

Crude watery extracts of five medicinal plants were administered orally by the help of a dropper. The treatment was repeated twice daily (morning and evening) for 12 days at the same dose and route. After treatment with crude watery extracts of five different medicinal plants all the treated chicks were observed carefully for 12 days and following parameters were studied.

Clinical examination:The time of disappearance of clinical signs and symptoms of coccidiosis and mortality if any was recorded carefully in the chicks treated with different medicinal watery extract.

Faecalsample examination:Faecal samples were obtained from all the treated 4 chicks of each group on every alternate day. At 9.00 a.m. in order to study the presence/absence of oocysts after treatment of chicks with watery extract of medicinal plants. Fecal sample was taken 1 gm from each group of chicks.

C. Comparative efficacy of chutra and thankuni leaves with that of Embazine

Sixteen infected (coccidiosis) chicks were used for this study. All the chicks were randomly divided into 4 groups each consisting of 4 chicks. The chicks and were kept in 4 different cages group wise and numbered as I to IV. Group I chicks was kept as control without giving any treatment. Group II and III chicks were given crude watery extract of chutra and thankuni leaves respectively at the dose rate of 10 ml/kg body weight. The dose (10 ml/kg) of crude watery extract was found to be most effective in treating the chicks infected with coccidiosis. On the other hand, group IV chicks received Embazine at the dose rate of 15 ml/gallon of drinking water and was supplied continuously for 3 days and fresh water was given for two days following Embazine again for 3 consecutive days.After administration of crude watery extracts and Embazine to different groups of chicks as mentioned above,all the treated chicks were kept under close observation for a period of 12 days.

RESULTS AND DISCUSSION

Ninety six chicks experimentally infected with coccidiosis were used in the present study with the objectives of 1) determining the effects of crude watery extract of mango, pineapple, guava, chutra leaves and whole of the thankuni and 2) the evaluating the comparative efficacy of crude watery extract of chutra leaves and whole of the thankuni with that of patent modern drug Embazine against coccidiosis in chicks.

The crude watery extract of mango leaves in doses of 1ml, 5ml and 10ml/kg body weight significantly decreased the oocyst count in groups II to IV chicks respectively. The maximum reduction to the extent of 64-and 36 percent was observed following 1 and 5 ml/kg body weight respectively. The reduction of oocyst was maximum on day 12 day following 10ml/kg body weight. In group 11, three chicks died within 4-6 days of treatment. The rest chick of this group collapsed within 8 days of medication. In group III, all the chicks died within 2-6 days of treatment. Ultimately none of the chicks of these two groups survived following 1 and 5 ml/kg body weight crude watery extract of mango leaves. In group IV, three chicks died within 10 days of medication and only one chick of this group(receiving 10ml/kg extract) survived.

Following 1,5 and 10ml/kg body weight crude watery extracts of pine apple leaves, oocyst count reduced significantly to the extent of 45, 57 and 88 percent in groups VI, VII and VIII chicks respectively. In group VI chicks 3 chicks died within 2-4 days after treatment with 1 ml/kg body weight of crude watery extract of pine apple leaves. One chick of this group survived throughout the study period. After administration of 5ml/kg body weight of pine apple extracts all the chicks died with 8 days of treatment. Whereas 2 chick died (within 2-8day) and 2 chick survived following 10 ml/kg body weight of pine apple extract.

Significant decrease of oocyst count to the extent of 9-44 percent was observed following administration of higher doses i.e. 5ml, 10ml/kg body weight of crude watery extracts of guava leaves against coccidiosis in groups XI and XII chicks whereas oocyst count was not decreased significantly after administration of 1ml/kg of guava extract in group X and XI chicks.
All the chicks of groups X and XI receiving 1ml and 5ml/kg body weight crude watery extracts of guava leaves died within 8 days of medication. In group XII three chicks died within 10 days and only one chick survived in this group.

In group XVI chicks, no oocyst was found on day 8 and all the chicks survived following administration of 10ml/kg body weight of crude watery extract of chutra leaves. Oocysts disappeared in group XIV and XV chicks within 8-10 days after treatment with 1ml and 5ml/kg body weight of chutra leaves extract. One chick in each group of XIV and XV died within 4-6 days following administration of 1 and 5 ml/kg body weight chutra leaves extract. Other 3 chicks of these groups survived during the study period.

No oocysts were observed after 8 days in groups XVIII, XIV and XX chicks following administration of 1,5 and 10 ml/kg body weight of crude watery extracts of whole of thankuni. However one chick in each group of XIX and XX died within 4-6 days after administration of 5 and 10 ml/kg body weight of thankuni extracts. Two of 4 chicks died within 4-6 days after administration of 1ml/kg body weight of thankuni extracts in group XVIII chicks.

Sixteen chicks were used for comparative study. They were divided into 4 groups each comprising of 4 chicks. Group I chicks was kept as control giving no treatment. The administration of crude watery extracts of whole of thankuni to group II chicks could not save all of the chicks. One chick of this group died within 4 days of medication, other 3 chicks of this group survived and become normal. The oocyst count also decreased significantly and found to be absent in faecal sample of surviving chicks on 12 day of treatment.

In group III chicks no oocyst was observed after 6 days of treatment with crude watery extract of chutra leaves and all the chicks of this group remained alive during the study period. Similarly, administration of Embazine in dose rate of 15 ml/gallon water to group IV chicks afforded complete protection against coccidiosis and oocyst was disappeared in the faecal sample after 4 days of treatment.

Clinical symptoms of coccidiosis i.e. blood mixed faeces in droppings, inappetance, ruffled feather, drowsiness etc. started to disappear from 3-4 days onwards in all the surviving chicks following administration of crude watery extract of different medicinal plants.

Post mortem examination was performed in the dead chick of all groups. Accumulations of blood in ceca and cecalcores, free blood in lumen of intestine were observed. The ceca become dark red in colour and firmer than normal. Necrotic mass of yellowish, cheesy blood stained debris were also observed. From the above findings it may be noted that some herbal extracts have anticoccidial activities. Almost similar types of findings were reported by many investigators (Mathis *et al.* 1984; Kirtikar and Basu, 1980; Siddique, 1961). Extracts of chutra leaves was more effective than other extracts (mango, guava, pineapple, thankuni).Due to unavailability of literature, the findings not properly explained in this study. In near furure the investigation needs to be performed to demonstrate the active ingredients of chutra having anticoccidial activity.

REFERENCES

1. Arsad MA, Khan AH and Zaman MB, 1956. Anote on the plants of medicinal values found in Pakistan. Medicinal Plant Branch, Pak. Forest Research Institute Abottabad.

2. Dhama K, Latheef KS, Mani S, Samad HA, Karthik K, Tiwari R, Khan RU, Alagawany M, Farag MR, Alam, GM, Laudadio V and Tufarelli V, 2015. Multiple Beneficial Applications and Modes of Action of Herbs in Poultry Health and Production-A Review. International Journal of Pharmacology, 11: 152-176.

3. FAO, 1980. Preliminary Study on Traditioal Systems of Veterinary Medicine. FAO regular programme, No. RAPA43. p 49-58.

4. Fernando ST, 1957. Coccidia of Goats in Ceylon - Preliminary Observations. Ceylon Veterinary Journal 1-2: 19-23.

5. Fish FF, 1932. Some factors in the control of coccidiosis of poultry. Journal of American Veterinary Medicine, Ass. 80: 543-559.

6. Gill BS, 1954. Speciation and Viability of Poultry Coccidia in 120 fecal samples preserved in 2.5% K-dichromate Solution. Indian Journal of Veterinary Science, 24: 245-247.

7. Harlen HDA and James AA, 1978. First lesson in duck raising. IVS Bulletin No-2, Dhaka, Bangladesh, pp. 17-18.

8. Kirtikar and Base, 1980. Indian Medicinal plants. Publisher: Bishen Singh Mahendra Pal Singh, Vol. 1 Edn-2, pp. 652-653.

9. Mathis GF McDougald LR, Mc Murray B, 1984. Effectiveness of therapeutic Anticoccidial drugs against recently isolated coccidia. Poultry Science, 63: 1149-53.

10. McDougald LR, 2003. Coccidiosis. In: Disease of Poultry (eds) Y. M. Saif, H. J. Barnes, J. R. Glisson, A. M. Fadly, L.R. McDougald, and D. E. Swayne. Iowa State Press, Ames, Iowa, USA. pp. 974-991.

11. Siddique S, 1961. International Symposium of Medicinal Plants of the Middle East. Pakistan Journal of Science, 13: 6-11.

12. Soltys A, 1970. Effects of some of external environmental factors on the infectivity of Eimeriatenellaoocystswiad. Parazyt,16: 175-181.

8

ESTABLISHMENT OF HEALTH MANAGEMENT PACKAGE FOR NATIVE SHEEP OF BANGLADESH

KBM Saiful Islam[1], Md. Ershaduzzaman[2], Md. Nuruzzaman Munsi[2], Md. Humayun Kabir[3], Sompa Das[3] and Md. Hazzaz Bin Kabir[4]

Department of Medicine and Public Health, Sher-e-Bangla Agricultural University, Dhaka, Bangladesh; [2]Goat and Sheep Production Research Division, Bangladesh Livestock Research Institute, Savar, Dhaka 1341, Bangladesh; [3]Conservation and Improvement of Native Sheep through Community Farming and Commercial Farming, Bangladesh Livestock Research Institute, Savar, Dhaka 1341, Bangladesh; [4]Department of Microbiology and Parasitology, Sher-e-Bangla Agricultural University, Dhaka, Bangladesh

Corresponding author: KBM Saiful Islam: E-mail: vetkbm@vahoo.com

ARTICLE INFO	ABSTRACT

Key words
Coccidiosis
Phytotherapy
Chicken

The present study was designed to explore the disease status of sheep in Bangladesh with the aim of developing health management package for sheep for better and efficient sheep production in Bangladesh. Both retrospective and prospective investigations on the incidence and prevalence of different diseases of sheep have been conducted in the study areas. Information on the disease related factors like health status, sex and age, vaccination, deworming, etc. were collected. Samples (faeces, blood, etc.) were collected from the diseased/dead animals and preserved following the standard procedure. Faecal samples were examined to determine the parasitic loads and faecal as well as blood samples were examined to determine any protozoan infection present in the study animals. Samples from diarrhoeic animals were studied to isolate and identify causal organisms. Antibiotic resistance and sensitivity studies of the aetiological agents responsible for common infectious diseases in sheep were also conducted in order to suggest the most suitable antibiotic to treat the concerned diseases in the field. Since helminthic infections, diarrhea, pneumonia and foot rot were found to be the mostly occurring health hazards in sheep of all ages, especial attempts were made to develop herbal based novel approaches to treat and control major intestinal helminthic infestations of sheep. However, gentamicin and ciprofloxacin were found most effective antibiotics and recommended to use in diarrhoeal cases in the field. On the other hand, methanol extract of mahogany seeds (100 mg), betel leaf (100 mg) and dodder (100 mg) were found significantly effective against 100% worms in 2 hours *in vitro* and thus recommended to be used in the field against helminthic infection in sheep. Therefore, a health management package for native sheep of Bangladesh can be designed using the present findings with some fluctuations for different sites after doing the *in vivo* evaluation of the medicinal plants used in this experiment.

INTRODUCTION

The small ruminants (goat and sheep) play very important role in the rural economy of Bangladesh by providing source of employment, women empowerment and the tool for poverty alleviation. Goat production has drawn the attention of the policy makers over a decade back and become popular to the livestock farmers especially the small scale livestock farmers of Bangladesh. Nowadays, especial emphasis has also been given to the preservation, conservation and rearing of native sheep through commercial and community farming. Considering the importance and potentiality of sheep production, the government of Bangladesh has granted the second phase of the developmental project 'Conservation and Improvement of Native Sheep through Community Farming and Commercial Farming (CINSCFCF)'. But numerous health problems of sheep continue to be the major threat to the efficient sheep production in Bangladesh. Among the multitude of problems hindering the development of sheep industry in Bangladesh, infectious diseases and endoparasitism constitute serious threats to the successful small ruminants industry like sheep farming. The diseases are responsible for loss of production, reduced fertility rate, reduced feed conversion efficiency, higher production cost, higher risk of zoonotic diseases and public health hazards. All these diseases lead to a great economic loss to the farmers. For these reasons, planned animal heath and good herd management packages designed to maintain the optimum animal heath and production are necessary for better sheep health management. Therefore, the present study was proposed with the aim of developing health management packages for native sheep for better and efficient sheep production in Bangladesh.

MATERIALS AND METHODS

The study was conducted in 11 selected areas of Bangladesh where the CINSCFCF Project is being implemented. The study has been proposed for a period of two years (July 2013 to June 2015) and the first phase (July 2013 to June 2014) of the study has been designed to explore the disease status of sheep in Bangladesh. Both retrospective and prospective investigations on the incidence and prevalence of different diseases of sheep have been conducted in the study areas. Information on the disease related factors like health status, sex and age, vaccination, deworming, etc. were collected. Samples (faeces, blood) were collected from the diseased/dead animal and preserved following the standard procedure.

Faecal samples were examined to determine the parasitic loads and fecal as well as blood samples were examined to determine any protozoan infection present in the study animals. Samples from diarrhoeic animal were studied to isolate and identify causal organism. Antibiotic resistance and sensitivity studies of the aetiological agents responsible for common infectious diseases in sheep were also conducted in order to suggest the most suitable antibiotic to treat the concerned diseases in the field. Since helminthic infections, diarrhea, pneumonia and foot rot were found to be the mostly occurring health hazards in sheep of all ages, especial attempts were made to develop herbal based novel approaches to treat and control major intestinal helminthic infestations of sheep. Locally available herbal plants known to have anthelmintic properties like neem (leaves), mahogany (seeds), betel (leaves), dodder (swarnalata; whole plant), bitter gourd (fruit) were collected and both aqueous and solvent extract were prepared. Mature live stomach worms were collected from sheep slaughtered at the abattoir to determine the effect of aqueous or methanolic extract on the parasites. A minimum of ten worms were exposed in three replicates to different concentrations of the plant extracts of each plant.

RESULTS AND DISCUSSION

The GIT parasitic infection, diarrhoea, pneumonia, PPR and foot rot are the major disease problems of sheep population in Bangladesh (Figure 1). A total of 334 sheep were examined for helminthic infection of which 299 (89.5%) were found infected with one or more species of helminth parasites, among them *Fasciola gigantica, Schistosoma indicum, Paramphistomum cervi, Cotylophoron cotylophorum, Moniezia expansa, Haemonchus contortus, Oesophagostomum sp. Trichuris ovis, T. vitulorum* and *Ostertagia sp.* were more prevalent. But in a study at Tangail district of Bangladesh, 81.1% sheep were found positive for one or more species of helminth parasites (Sangma *et al.,* 2012). In another study at Mymensingh district of Bangladesh, it was found that 94.67% sheep were found positive for one/more species of helminth parasites (Mazid *et al.,* 2006). These differences in findings may be due to geographical and seasonal variations.

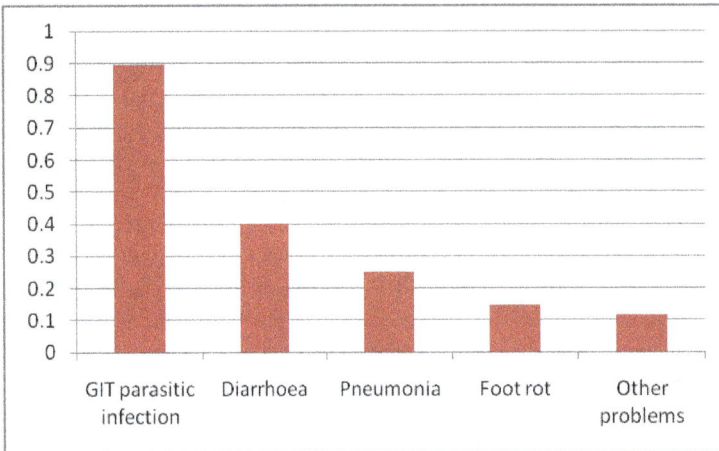

Figure 1. Status of disease incidence in all age group of sheep

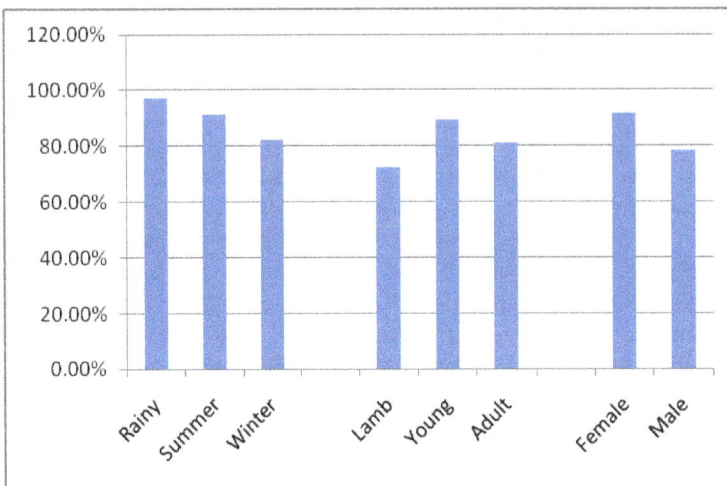

Figure 2. Status of parasitic infection in relation to season, age and sex in sheep

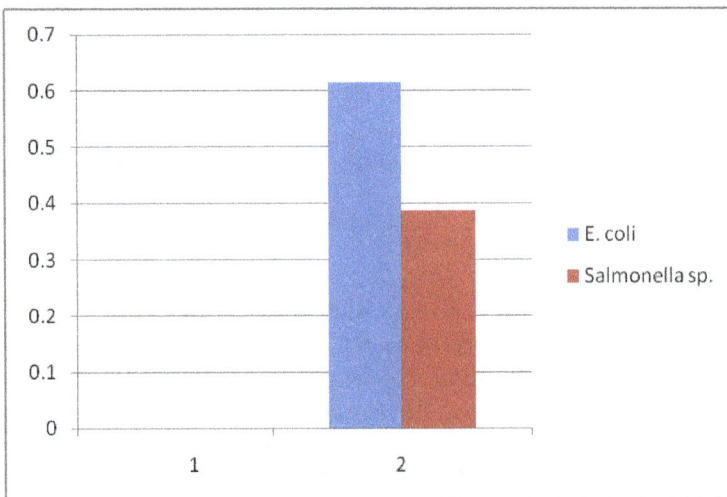

Figure 3. Status of bacterial infection in relation to diarrhoea in sheep

Relatively higher occurrence of parasitic infection in sheep was recorded in rainy season (97.41%) followed by summer (91.33%) and winter (82.35%). Prevalence of helminths in sheep was significantly higher in young sheep (89.27%) than in adult (81.11%) and in lamb (72.41%). Female (91.61%) sheep were found more susceptible than their male counterpart (78.24%) (Figure 2). However, parasitic infection is a matter of great concern for the livestock owners as it significantly interferes with their productivity. Heavy burdens of trichostrongylid nematodes can lead to anaemia, oedema, haemorrhage, traumatic injury to the intestinal epithelium and even death. In hyperacute cases of haemonchosis, sheep died suddenly from haemorrhagic gastritis (Urquhart et al., 1996). In addition to losses through mortality, major losses are attributed to reduced feed efficiency, lowered production of meat, and labor and drugs associated with control (Hartwig, 2000).

Bacteriological investigations of diarrhoeic cases revealed E. coli and Salmonella sp. as the aetiological agents in 61.44% and 38.56% cases, respectively (Figure 3). In vitro drug sensitivity profile of E. coli and Salmonella isolates against a set of common antibiotics indicated a marked heterogeneity in the sensitivity pattern. But they were found sensitive to gentamicin and ciprofloxacin. Thus, these antibiotics are recommended to use in diarrhoeal cases in the field.

The study revealed that the herbal plants used in this experiment have remarkable anthelmintic properties. Herbal extracts exhibited anthelmintic effects in a dose dependent manner and all the worms were found dead within 6 hr post-exposure. Mahogany (seeds) (100 mg), betel leaf (100 mg) and dodder (100 mg) were found significantly effective against 100% worms in 2 hours in vitro (Table 1). The anthelmintic properties of these herbal plants were proved by Sharma et al. (2003), Chandrawathani et al. (2006), Islam et al. (2008), Adate et al. (2012), Priscilla et al. (2014) and Akter et al. (2014). Thus, these herbal plants are recommended to be used in the field against helminthic infection in sheep.

Table 1. In vitro anthelmintic efficacy of different concentrations of methanol extracts of selected indigenous medicinal plants against adult gastrointestinal nematodes of sheep

Medicinal plants	Effectiveness (%)		
	25 mg	50 mg	100 mg
Mahogany (seeds)	40%	70%	100%
Betel (leaves)	20%	60%	100%
Dodder (whole plants)	30%	50%	100%
Bitter gourd (fruit)	40%	60%	100%
Neem (leaves)	40%	70%	90%

CONCLUSION

As helminthic infections, diarrhea, pneumonia and foot rot are the mostly occurring health hazards in sheep of all ages, in addition to hygienic measures regular deworming practice should be adopted by the farmers. Regarding antibiotic sensitivity, gentamicin and ciprofloxacin were found most effective and recommended to use in diarrhoeal cases in the field. On the other hand, mahogany (100 mg), betel leaf (100 mg) and dodder (100 mg) were found significantly effective against 100% worms in 2 hours in vitro and thus recommended to be used in the field against helminthic infection in sheep. Therefore, a health management package for native sheep of Bangladesh can be designed using the present findings with some fluctuations for different sites after doing the in vivo evaluation of the medicinal plants used in this experiment.

REFERENCES

1. Adate PS, Parmesawaran DS and ChauhanY, 2012. In vitro Anthelmintic activity of stem extracts of Piper betle Linn against Pheritima posthuma. Pharmacology Jiurnal, 4: 61–65.
2. Akter KN, Karmakar P, Das A, Anonna SN, Shoma SA and Sattar MM, 2014. Evaluation of antibacterial and anthelmintic activities with total phenolic contents of Piper betel leaves. Avicenna Journal of Phytomedicine, 4: 320–329.

3. Chandrawathani P, Chang KW, Nurulaini R, Waller PJ, Adnan M, Zaini CM, Jamnah O, Khadijah S and Vincent N, 2006. Daily feeding of fresh Neem leaves (A.indica) for worm control in sheep. Tropical Biomedicine, 23: 23-30.

4. Hartwig N, 2000: Sheep health- fact sheet no. 8: Control of internal parasites of sheep. USDA, Iwoa State University.

5. Islam KR, Farjana T, Begum N and Mondal MMH, 2008: In vitro efficacy of some indigenous plants on the inhibition of development of eggs of *Ascaridia galli* (Digenia:Nematoda). Bangl. J. Vet. Med., 6: 159–167.

6. Mazid MA, Bhattacharjee J, Begum N and Rahman MH, 2006: Helminth parasites of the digestive system of sheep in Mymensingh, Bangladesh. Bangladesh Journal of Veterinary Medicine, 4: 117-122.

7. Priscilla FX, Amin MR and Rahman S, 2014. Comparative study of neem (*Azadirachta indica*), bitter gourd (*Momordica charantia*) extract as herbal anthelmintic and albendazole as chemical anthelmintic in controlling gastrointestinal nematodes in goats. IOSR Journal of Agriculture and Veterinary Science, 7: 33-37.

8. Sangma A, Begum N, Roy BC and Gani MO, 2012: Prevalence of helminth parasites in sheep (*Ovis aries*) in Tangail district, Bangladesh. Journal of Bangladesh Agricultural University, 10: 235-244.

9. Sharma V, Walia S, Kumar J, Nair MG and Parmar BS, 2003. An efficient method for the purification and characterization of nematicidal azadirachtins A, B, and H, using MPLC and ESIMS. Journal of Plant Physiology, 160: 557-564.

10. Urquhart GM, Armour J, Duncan JL, Dunn AM and Jennings FW, 1996: Veterinary Helminthology. In: Veterinary Parasitology, 2nd edn., Blackwell Science Ltd., p 21.

9

STUDIES ON PREVALENCE OF ASCARIASIS IN INDIGENOUS CHICKENS IN GAIBANDHA DISTRICT AND TREATMENT BY PINEAPPLE LEAVES EXTRACT

Mst. Kamrunnaher Akter[1], ABM Jalal Uddin[1], Maksudur Rashid[1], Fahima Binthe Aziz[2], Md. Bazlar Rashid[2] and Mahmudul Hasan[2]

[1]Department of Livestock Services (DLS), Bangladesh; [2]Department of Physiology and Pharmacology, Hajee Mohammad Danesh Science and Technology University, Dinajpur-5200, Bangladesh

*Corresponding author: ABM Jalal Uddin; E mail: abulbasar47@gmail.com

ARTICLE INFO	ABSTRACT
Key words Ascariasis Chickens Antiparasitic Pineapple leaves Efficacy	Ascariasis is very common in indigenous chicken causing retarded growth, low productivity and mortality. Considering the problem of *Ascaridia galli* in chickens, anthelmintic resistance, high cost and human health hazard of chemical anthelmintic the use of medicinal plant is an alternative choice. The study was conducted to determine the incidence of ascariasis in Polashbari upazilla of Gobindhaganj district during July to November 2012 and subsequently evaluated the efficacy of pineapple (*Ananus comosus*) leaves extract against ascariasis infected chickens. Out of 500 chickens examined for presence of *A. galli* infestation by faecal sample examination, 365 hens and 135 cocks. The 292 female (80%) and 119 male (88.15%) were found infected with *A. galli*. The highest infection rate 95.26% was found in 60 to 90 days of age group. Infected chickens were treated with pineapple leaves extract @ 1ml/kg body weight per OS for 7 consecutive days. The efficacy of anthelmintic treatment was evaluated by counting fecal egg per gram (EPG) compared with pretreatment values. Body weight and hematological changes of each chicken was recorded in pre and post treatment. In the untreated control chickens the average EPG increased from 300 ± 11.07 to 340 ± 13.96. The average EPG reduced from 300 ± 11.07 to 60 ± 7.40 within 28 days of pineapple treatment. The mean body weight gain in treated chicken was significantly ($p<0.01$) higher than the control. Pineapple leaves extract increased the TEC, Hb and PCV and decreased TLC and ESR values of chickens. But in control group TEC, Hb and PCV decreased and TLC and ESR values increased. It may be concluded that pineapple leaves extract treatment effectively reduced the ascariasis load in chicken and improved body weight.

INTRODUCTION

Poultry is a promising sector in Bangladesh which is increasing day by day. Poultry production is hindered by many problems among which various diseases namely bacterial, viral and parasitic infections are the most important (Ojok, 1993). In fact poultry of Bangladesh are infested with various parasites (Sarkar, 1976). Management system plays an important role in the occurrence of parasitic disease. The parasitic load leads to lower productivity, retarded growth rate and death of birds (Barger, 1982; Sykes, 1994). Ascariasis caused by *Ascaridia galli* is a common parasitic problem of chicken both in rural and farm conditions in Bangladesh (Islam and Shaikh, 1967;Haq, 1986). *A. galli* causes extensive economic losses in different ways such as loss of weight gain, meat production, egg production and death of birds (Kamal, 1989). Piperazine citrate is widely used for the treatment of ascariasis in chickens. The use of anthelmintics by farmers for poultry parasite is not usual and strategic. Frequent and improper use of anthelmintics increases the resistant population of nematodes (Waller *et al.*, 1987). Furthermore the withdrawal period of anthelmintic is not maitainded for consumption of poultry in Bangladesh which can cause significant health threat. Therefore, the use of safe and cost effective alternative approach for treatment of poultry parasite is necessary. There are several indigenous medicinal plants have anthelmintics action and used against both ecto and enodparasites in Bangladesh (Mostofa, 1983; Mannan *et al.*, 1997). Pharmacological actions along with therapeutic trial of these plants may be studied experimentally, which might prove worthy of medical value. Pineapple leave extract is used as anti-inflammatory, anti bacterial and anthelmintic agent (Mostofa, 1983 and Amin *et al.*, 2009). Considering the diversified pharmacological function of pineapple the present study was designed to evaluate the efficacy of pineapple leave extract on ascariasis in indigenous chicken.

MATERIALS AND METHODS

The experiment was conducted at the Department of Physiology and Pharmacology, Hajee Mohammed Danesh Science and Technology University, Dinajpur and Upazila Livestock Office, Palashbari, Gaibandha, during the period from July to November, 2012

Study area and sampling

A total of 500 native chickens (*Gallus gallus domesticus*) of 2 to 7 months age from different villages of Palashbari Upazila of Gaibandha District were examined to study the prevalence according to sex, age and location. Fecal samples were collected from the cloaca during early in the morning. The sample were packed within polythene bags and sent as soon as possible to the laboratory for examination. Direct smear method and Stoll's ova counting technique were used for fecal sample examination following the procedure described by Urquhart (2003) and Soulsby (1982).

Experimental bird

Fourty five indigenous chickens having infection with *Ascaridia galli* were selected for this experiment. The chickens were allowed to take rest for 7 days for adaptation. The experiment was carried out in upazila livestock office, Palashbari, Gaibandha. The age and body weight of chickens ranged from 2 to 7 months and 300 to 500 gm, respectively. The chickens were supplied with normal diet and water.

Fecal eggs count

For determination of infectivity, fecal samples were collected and eggs were counted by Stoll's ova counting technique and direct smear method through microscopy following the procedure described by Rahman et al. (1996) and Soulsby (1982). At least three slides from each faecal sample were examined.

Drug and plant

Pineapple leaves (*Ananus comosus*) were collected from the horticulture garden of the University. Ten percent (10%) water extract of pineapple leaves was prepared freshly, 20 gm pineapple leaves was grinded in mortar and pastle, the extract was made in 20ml water. The extract was then administered 1ml/kg orally by dropper.

Experimental design

All the 30 chickens randomly divided into 2 groups (control and treatment). Control group was without treatment and treatment group was treated with pineapple leaves extract. Pineapple leaves extract was administered orally @ 1 ml/kg bwt by dropper by 7 consecutive days. All the chickens of treated and control groups were closely observed for 28 days after treatment and on body weight, feeding efficiency, feather coat.

Hematological examination

Blood samples were collected from the wing vein of chicken of both control and treated groups at pre-feeding and during feeding (28 days) period at 7 days interval. Total erythrocytes count (TEC), Erythrocytes sedimentation rate (ESR), Packed cell volume (PCV) and Total leukocyte count (TLC) were performed as per methods described by Schalm *et al.* (1975). Hemoglobin estimation was performed as per method described by Coffin (1955).

Postmortem examination

Before treatment three chickens from each group were also slaughtered to count the number of parasites (Ascarids) and to see if there were any pathological changes present. After treatment three chickens from each group were slaughtered to count number of parasites (*A. galli*) and to see if there were any pathological changes present on 14th and 28th day of treatment.

Statistical analysis

Comparison of the mean values of the treatment against those of the control was performed by Student's t-test and the level of probability considered significant when $p < 0.05$.

RESULTS AND DISCUSSION

Out of 500 chickens 365 were hens and 135 were cocks. Among the female 292 (80%) and male 119 (88.15%) birds were found infected with *A. galli*.

Prevalence of ascariasis

The prevalence of ascariasis in chickens indifferent villages of Palashbari upazila of Gaibandha district is presented in table 1. The highest (93%) and lowest (74.74%) prevalence of ascariasis were recorded in village Andua and Nuniagari, respectively.

Table 1. Incidence of ascariasis in chickens of different age groups of Palashbari upazila of Gaibandha District

Category	Category-1 (2-3 months)	Category-2 (above 3-5 months)	Category-3 (above 5-7 months)	Total
Number of chicken examined	190	180	130	500
Number of infected chicken examined	181	167	72	420
Rate of infection (%)	95.26	92.78	55.38	84

A significant higher incidence (95.26%) of the infection was recorded in chickens between 2-3 months of age, followed by 92.78% and 55.38% between age group of 3-5 months and 5-7months respectively. Similar findings have reported by Sarker *et al.* (2009), Gauly *et al.* (2005), Romanenko *et al.* (1985). The results indicate that the rearing or management system in different village and age factors played an important role in ascariasis in chicken.

Effect on body weight

The effect of pineapple leaves on body weight was observed for 28 days at 7 days interval. Mean body weight of each group of chickens at pretreatment and post treatment period is presented in table 2.

Table 2. Effects of pineapple leave extract on body weight (gm) in chicken at different treatment period.

Group of chicken	0 day	7th day	14th day	21st day	28th day	Mean weight gain (%)
Control	402.466±3.65	398.241± 3.88	405.53± 3.58	403.413 ±3.54	406.06±3.46	0.89
Pineapple leaves extract @ 1gm/kg bwt orally	430.03± 5.29	431.99± 5.92	437.79± 5.19	439.07 ± 5.56*	446.09*±5.26	3.74

Values given above the represent the mean ± SE of 5 chickens

Table 3. Effects of pineapple leave extract on fecal egg count.

Group of Chickens	0 day	14th day	28th day
Control	300 ± 11.07	320 ±11.42	340 ± 13.96
Pineapple leaves extract@ 1gm/kg bwt orally	300 ± 11.07	140 ± 7.40	60 ± 7.40

Values given above the represent the mean ± SE of 5 chickens

Effect on parasite

Pineapple leaves extract against ascariasis showed reduction of EPG count on 14th and 28th day in the treated group of chickens whereas EPG count was increased in control group (Table 3). The data showed that pineapple leaves extract was found to be about 53% effective within 14 days of treatment and 80% effective within 28 days of treatment. These results are in agreement with earlier reports of Patra *et al.* (2010), Amin *et al.* (2009), Sujon *et al.* (2008), Islam *et al.* (2005), Khalid *et al.* (2005) and Khatun *et al.* (1995).

Table 4. Effects of pineapple leave extract on number of parasites in chickens.

Group of chicken	Pre-treatment (No.)	After drug administration (post-treatment) (No.)	
	0 day	14th day	28th day
Pineapple leaves extract @ 1gm/kg bwt	8±1.63	3±1.32	1±1.31
Control	10±1.41	14±1.41	15±1.90

Values given above the represent the mean ± SE of 3 chickens

Effect on hematological parameters

On day '0' the mean value of initial body weight of treated group was 430.03±5.29 gm and on the 28th day of post treatment, the mean value of body weight was 446.096±5.26 gm. The body weight was significantly (p<0.05) increased (3.74% on 28th) at post treatment period compared to pretreatment. But percentage of body weight gain of the control group was very negligible as 0.89% on 28thday. The present findings support the earlier observation of Hoque *et al.* (2006), Khalid *et al.* (2005), Islam et al. (2004, 2005), Khatun *et al.* (1995).

Table 5. Effects of pineapple leave extract on TEC (million/cu mm) in indigenous chickens in different treatment period

Blood Parameters	Group	0 day	7th day	14th day	21st day	28th day
TEC (million/ cu mm)	Pineapple Treated	3.53±0.26	3.53±0.26	3.56±0.265	3.61±0.29	3.7±0.13
	Control	3.20±0.32	3.17±0.46	3.056±0.34	3.12±0.30	2.96±0.19
Hemoglobin (gm%)	Pineapple Treated	9.48±0.54	9.52±0.47	9.6±0.38	9.6±0.35	9.8±0.53
	Control	9.14±0.68	8.96±0.57	8.9±0.50	8.88±0.51	8.82±0.50
PCV (% 30 minutes)	Pineapple Treated	19.36±0.88	20.86±0.8 4	20.92±0.78	21.64±0.78	22.38±0.65
	Control	20.46±0.88	20.02±0.78	19.8±0.78	19.12±0.93	18.22±0.69
ESR (mm/1st hour)	Pineapple Treated	0.5±0.27	0.44±0.34	0.48±0.29	0.42±0.36	0.4±0.32
	Control	0.62±0.39	0.56±0.4	0.6±0.32	0.58±0.29	0.48±0.29
TLC (10^3/mm^3)	Pineapple Treated	7.66±0.64	7.60±0.45	7.56±.30	7.53±0.28	7.43±0.39
	Control	10.23±0.69	10.32±0.75	10.37±0.71	10.41±0.64	10.49±0.81

Values given above the represent the mean ± SE of 5 chickens

The administration of pineapple leaves extract increased the TEC, Hb and PCV and reduced TLC and ESR values of chicken. But in non treated control group, TEC, Hb and PCV were decreased and TLC and ESR values were increased. The maximum reduction or increased values were observed at 28thday of post treatment. Changes in hematological parameters were similarly reported by other researchers (Hoque *et al.*, 2006; Islam *et al.*, 2005; Khatun *et al.*, 1995).

Postmortem examination

There was no significant pathological change in any internal organs of the chicken of the treated groups. Reduction of parasite count was found on 14th and 28th day in the group of chicken treated with pineapple extract. The highest reduction of number of parasites was recorded on 28th day of treatment. On the other hand, number of parasites was increased day by day in control group A. This findings support the earlier observation made by Gauly *et al.* (2005) and Malakhov (1988).

Effect on feeding efficiency

Treatment of chicken with pineapple leaves extract significantly increased feeding efficiency of the chickens. But feeding efficiency decreased gradually in chickens of control group.

Effect on feather coat

The feather coats of all treated chicken were observed smooth and shiny at 28th day of Post treatment period. Rough and discoloured feather coats obverted in non treated control were due to severe parasitic infestation. *A. galli* is prevalent in indigenous chicken in different villages of Gaibandha district which suggest treating the infested chicken with effective anthelmintics and implementing regulars deworming program of chicken in the study areas. Pineapple leaves showed highly encouraging efficacy as an anthelmintics. Considering the availability and low cost pineapple leaves can be used to treat chicken ascariasis. Since this study is a preliminary work in a small population of chicken further studies must be carried out to evaluate the adverse effects, bio-chemical analysis of pineapple leaves against ascariasis in chickens.

REFERENCES

1. Amin M R, Mostofa M, Hoque ME and Sayed MA, 2009. In vitro anthelmintic efficacy of some indigenous medicinal plants against gastrointestinal nematodes of cattle. Journal of Bangladesh Agriculture University, 7: 57–61.

2. Barger IA, 1982. Helminth parasites and animal production. In: Symons, L.E.A., A.D. Donald and J.K. Dineen (Eds), Biology and control of endoparasites. Academic press, Sydney. pp : 133-155.

3. Coffin DL (1955). Manual of Vterinary Clinical Pathology (3rd edn). Comstock Publishing Association, Inc. New York, USA.

4. Gauly M, Homannb T and G Erhardt, 2005. Agerelated differences of Ascaridia galli egg output and worm burden in chickens following a single dose infection. Veterinary Parasitology, 128(1-2): 141-148.

5. Haq MS, 1986. Studies on helminthes infections of chicken under rural condition of Bangladesh. Bangladesh Veterinary Journal, 20: 55-60.

6. Hoque M E, Mostofa M, Awal MA, Choudhury ME, Hossain MA and Alam MA, 2006. Comparative efficacy of piperazine citrate, levamisole and pineapple leaves extract against naturally infected ascariasis in ndigenous chickens. Bangladesh Journal of Veterinary Medicine, 4: 27–29.

7. Islam SA, Rahman MM, Hossain MA, Chowdhury MGA and Mostafa M, 2005. Comparative efficacy of some modern anthelmintics and pineapple leaves with their effects on certain blood parameters and body weight gain in calves infected with ascarid parasites. Bangladesh Journal of Veterinary Medicine, 3: 33-37.

8. Islam AWMS and Shaikh H, 1967. Efficacy of ivermectin (pour on formulation) against gastrointestinal nematodiasis and ectoparasites in cattle of Bangladesh. Indian Journal of Pharmacology. 31: 234-238.

9. Islam SA, Mostofa M, Awal MA and Khan KA, 2004. Efficacy of pineapple leaves extract compared with modern anthelmintics against ascariasis in calves. The Bangladesh Veterinarian, 21: 9-13.

10. Kamal AHM, 1989. Pathological investigation on the mortality of chickens in Bangladesh Agricultural University Poultry Farm. M.Sc. (Vet. Science) Thesis, Department of Pathology, Bangladesh Agricultural University, Mymensingh.

11. Khalid S MA, Amin MR, Mostofa M, Choudhury ME and Uddin B, 2005. Effects of indigenous medicinal plants (neem and pineapple) against gastrointestinal nematodiasis in sheep. International Journal of Pharmacology, 1: 185-189.

12. Khatun M, Awal MA, Mostofa M and Rashid MSH, 1995.Comparative efficacy of pineapple leaves with Fenbendazole against gastro-intestinal nematodes in goats. Bangladesh Veterinary Journal, 29: 75-78.

13. Malakhov Av, Korchagin AI, Pevneva VD and Nadykto MV, 1988. Tolerance of Ascaridia galli to piperazine. Veterinariya, Moscow, No. 7, 38-39.

14. Mannan MA, Rafiq K, Mostofa M and Hason Q, 1997. Comparative efficacy of Ivomec pour on, Neguvon Ointment and Neem-tobacco herbal preparation against naturally occurring hampsore Lesion in cattle. Bangladesh Veterinary Journal, 31: 119-122.

15. Mian R, 2009. Comparative efficacy of the selected indigenous medicinal plants with a patent drug levamisole against ascariasis in village poultry. Bangladesh Journal of Veterinary Medicine, 7: 320-324.

16. Mostofa M, 1983. Efficacy of some indigenous medicinal plants against gastrointestinal nematodiasis in cattle and their comparative activity with that of Nemafax. M.Sc (Vet. Science) Thesis. Department of Pharmacology, BAU.

17. Ojok L, 1993. Diseases as important factor affecting increased poultry production in Uganda. Der Tropenland Wirt Zeitschrift fur die landwirtschaft in din Tropen and Subtropen. 94: 37-44.

18. Patra G, Lyngdoh WM, Ali MA, Prava M and Chanu KV, 2010. Comparative Anthelmintic Efficacy of Pineapple and Neem Leaves in Broiler Chickens Experimentally Infected with Ascaridia galli. International Journal of Poultry Science. 9: 1120-1124.

19. Rahman MH, Ahmed S and Mondol MMH, 1996. Introduction to Helminth Parasites of Animals and Birds in Bangladesh. RS. Tahsina Mostofa (1stedn.). 16(18): 105-106.

20. Romanenko PT, Troenko TA, Zaremba I A and Kuzyakin AV, 1985. Age variation in helminth infection on chicken's farm and factory faems in the Roostov region. Poultry Abstract. 11: 249-250.

21. Sarkar AJ, 1976. The prevalence of avian diseases in Bangladesh Agricultural University Farm. Bangladesh Veterinary Journal, 10: 61-66.Sarker RR, Mostofa M, Awal MA, Islam MS and

22. Sarker RR, Mostofa M, Awal MA, Islam MS and Mammon SA, 2009. Prevalence of ascariasis in village poultry at five upazilas under feni District. Bangladesh Journal of Veterinary Medicine, 7: 293-295.

23. Schalm DW, Jain NC and Carrol EJ, 1975. Veterinary Haematology. 3rd edn. Lea and Febiger Philadelphia. USA. p.622.

24. Soulsby EJL, 1982. Helminth Arthropods and Protozoa of Domesticated Animals. 7thedn, Bailliere Tindal and Cassell Ltd. London.

25. Steel JW and Symons LEA, 1982. Nitrogen metabolism in nematodiasis of sheep in relation to productivity. In: Symons, LEA, AD Donald and JK Dineen (Eds), Biology and control of endoparasites. Academic press, Sydney. pp: 235-256.

26. Sujon MA, Mostofa M, Jahan MS, Das AR and Rob S, 2008. Studies on medicinal plants against gastrointestinal nematodes of goats. Bangladesh Journal of Veterinary Medicine, 6: 179-183.

27. Sykes AR, 1994. Parasitism and Production in farm ruminanta. Animal Production, 59: 155-172.

28. Urquhart GM, Aremour J, Dunchan JL, Dunn AM and Jeninis FW, 2003. Veterinary Parasitological 2th edition, Blackwell sciences ltd.

29. Waller PJ, Asbakk K, Hrabok JT, Oksanen A and Nieminen M, 1987. Prolonged persistence of fecally excreted ivermectin from reindeer in a sub-arctic environment. Journal of Agriculture Food Chemistry, 54: 9112-9118.

10

STUDIES OF THE COMPARATIVE EFFICACY OF ALCOHOLIC EXTRACTS OF BIRONJA, TURMERIC, AND VERANDA LEAVES WITH THAT OF PATENT DRUG NILZAN AGAINST TREMATODIASIS IN SHEEP

Tahmina Begum, Bayzer Rahman, Sukumar Saha[1] Mahbub Mostofa and Md. Abdul Awal*

Department of Pharmacology, Faculty of Veterinary Science, Bangladesh Agricultural University, Mymensingh-2202, Bangladesh; [1]Department of Microbiology and Hygiene, Faculty of Veterinary Science, Bangladesh Agricultural University, Mymensingh-2202, Bangladesh

*Corresponding author: Md. Abdul Awal, E-mail: dmawalbau@gmail.com

---

ARTICLE INFO

Key words

Efficacy,
Trematodiasis,
Sheep,
Medicinal plants

ABSTRACT

The study compared the efficacy of some indigenous plants (Bironja, Turmeric and Veranda leaves) with that of patent drug Nilzan against Trematodiasis in sheep. Seventy five sheep suffering from trematode, aged 2-3 years and weighing about 10-12 kg. were used for this study. The sheep were divided into 15 equal groups A, B, C, D, E, F, G, H, I, J, K, L, M, N and O. Administration of three different doses of powdered Bironja 3,6 & 9 ml /kg body wt to the group A, B, C, D and E, Turmeric at the dose rate of 2,4 & 6 ml /kg body wt to the groups F, G, H, I and J, Veranda leaves 2, 4 and 6 ml /kg body wt to the groups K, L, M, N and O showed 15-52% , 11-15% and 4-8% efficacy, respectively against trematodes in sheep. Administration of Nilzan 30 mg/kg body wt orally was highly effective 73-89% against all the species of trematodes in sheep. Among medicinal plants Bironja was found to be most effective followed by turmeric. Veranda leaves were found to be totally ineffective against trematodiasis in sheep.

INTRODUCTION

There are 195.28 million of livestock in Bangladesh of which sheep 1.09 million (Amin, 1994).In Bangladesh sheep are raised for meat production, though they are basically duel purpose animal producing both meat and wool. Field veterinarians often speak of parasitic disease as being one of the important hindrances to sheep productivity in the country but very few published reports could be traced in respect of their clinical details. Due to various problems, the health, production and performance of livestock in Bangladesh are at the lowest ebb compared with those of other countries. Parasitism is also an important limiting factor of livestock production in most of the tropical and sub-tropical countries. It is established that infestations due to trematode undermine the health and productivity of animals. Asian Development Bank (ADB) report clearly mentioned the loss of productivity of animals in terms of mortality, loss of milk, meat, generation loss and loss of reproductive rate due to animal parasites to the extent of 50 % in Bangladesh (ADB,1984). Besides the use of various modern patent drugs for the treatment of various diseases people throughout the world have been using the traditional indigenous medicinal plants and herbs as remedial agents for prevention, mitigation and cure of disease conditions since long. A variety of medicinal preparations from indigenous herbs and plants are now-a-days manufactured by same pharmaceutical industries in India, Indonesia and Thailand. In our country the Hakms, Ayurveds and kabiraj's are using indigenous plants for the treatment of human being. In village condition animal diseases are treated by Kabirajs and Quacks who use various indigenous plants. Research in India towards the direction of application of indigenous plants and their products to veterinary practice have lead to the development of certain industries of which Indian Herbs Reasearch Company of Shaharanpur in Uttar Pradesh needs worth mentioning. Therefore, this study was conducted to evaluate the efficacy of Biranja, Turmeric and verenda against trematodiasis in sheep.

MATERIALS AND METHODS

Experimental animals
Eighty five sheep of indigenous breed were collected from adjacent area of BAU, Mymensingh. The sheep were allowed to graze on pasture of the BAU campus for about one month. These sheep were suspected to be suffering from helminth by observing clinical sign and symptoms. Microscopic examination of faeces of the suspected sheep was carried out for trematode egg count. Seventy five sheep were selected for this study on the basis of the physical and clinical examination and result of the fecal egg count. The age of sheep ranged between 2 and 3 years approximately. The weight of all selected sheep ranged between 10 and 12 kg.

Collection and preparation alcoholic extracts of Bironja, turmeric and veranda
Bironja and turmeric were purchased from local market. The veranda leaves were collected from Bangladesh Agricultural University campus. The collected sample were washed thoroughly with water, the veranda leaves and turmeric were cut into small pieces and sun dried for seven days. Then these samples were dried separately in a hot air woven at 45°C for six hours. The dried samples were taken into equal parts by measuring with balance and pulvarised to 60 meshes in a grinding machine separately.100 gm of each sample was stored in air tight bottle separately. Alcoholic extracts of selected medicinal plants were obtained by Soxlet method. The alcoholic extracts of each kind of sample was then kept in separate air tight bottle and labeled for further study.

Determinatio of biochemical parameters
Biochemical parameters like serum glucose and SGPT (serum glutamic pyruvic transaminase) were also determined by using autoanalyser(Model No.Reflotron M-06).

Drugs and chemical used
Nilzan(Tetramizol hydrochloride and oxyclozanide) was purchased from local market. Diagnostic kit(Glucose, SGPT) test combination used in this study were purchased from Fisons Bangladesh Ltd. Anticoagulant Haymes solution,0.14 hydrocloric acid solution, Wrights stain, 90% alcohol,10% formalin, normal saline (0.09) saturated salt solution, were prepared in the laboratory.

Statistical methods

The data was analyzed statistically between normal and treated values by the well-known students test (t" test).

RESULT AND DISCUSSION

Bironja: The oral administration of alcoholic extracts of Bironja in doses of 3, 6 and 9 ml/kg body wt. was partially effective against Fasciola spp. In sheep (table I) Nilzan was very effective against Fasciola spp. In group B (3 ml/kg) the mean EPG (Egg per gram) values were 910 at pre-treatment period 770 on 24th day after medication. Similarly, the mean EPG values was 920, before treatment and on day 24 after treatment the value was 690 with 6 ml/kg body wt. of Bironja in sheep. In group D (9 ml/kg) the mean EPG values were 900 and 640 respectively before treatment and on 24th day after treatment. Following Nilzan (30 mg /kg body wt) administration the mean EPG was 890 before treatment and 950 on 24th day of treatment. The administration of 3, 6 and 9 ml/kg of bironja extracts could reduce the mean EPG upto 15-31% against Fasciola spp. This result shows that Bironja was partial effective against Fasciola spp. In partial agrement of the present findings, Andraske et. al (1974) recorded 90% efficacy in cattle of Czechoslovakia and Afaz Uddin(1985) 93.18% in cattle of Savar Military farm, Dhaka.

In case of paramphistomum spp., in group B (3 ml/kg) the mean EPG count was 990 before treatment and on 24th day of treatment value was 860. In group C (6ml/kg) the mean EPG count was 990 before treatment and on 24th day of treatment the value was 580. Similarly, in group D, the mean EPG values were 1010 and 480 before treatment and on day 24 respectively of treatment with 9 ml alcoholic extract of Bironja per kg body weight. Following Nilzan administration the mean EPG was 980 before treatment on 24th day after treatment it was 130 in group E. In group A,which was kept as control, the mean EPG reduction was up to 13 to 52% against paramphistomum spp. The efficacy of highest dose (9 ml/kg) which reduced mean EPG upto 52% was satisfactory. The efficacy of Nilzan (30 gm/kg body wt) was also satisfactory (mean EPG reduction 87%).

Table 1. The comparative efficacy of Bironja with that of patent drug Nilzan against *Fasciola spp* in sheep.

Groups	Drug and dose/kg body wt	Pre-treatment period	Post-treatment period							% of EPG Reduction
			4th day	8th day	12th day	16th day	20th day	24th day	
		EPG	EPG	EPG	EPG	EPG	EPG	EPG	
A	Control	890± 22.36	910± 26.62	910± 28.32	920± 12.06	930± 11.05	940± 9.86	950± 10.09	
B	Bironja (3ml/kg)	910± 19.79	900± 17.32	890± 21.32	880± 11.79	850± 12.63	790**± 16.87	770**± 8.86	15.38 %
C	Bironja (6ml/kg)	920± 32.09	910± 8.86	890± 10.11	870± 19.31	790**± 7.32	740**± 8.38	690**± 3.79	25.00 %
D	Bironja (9ml/kg)	900± 22.02	890± 12.21	850± 17.71	790**± 7.59	740**± 2.08	650**± 6.75	640**± 10.90	31.11 %
E	Nilzan (30mg/kg)	890± 10.86	520**± 21.24	410**± 3.82	290**± 27.56	180**± 18.53	140**± 8.63	100**± 9.36	87.64 %

Values given above represent the mean ± SE of 5 sheep.
**Significant decrease (p<0.01).

Table 2. The comparative efficacy of Bironja with that of patent drug Nilzan against *Paramphistomum spp* in sheep.

Group of animal	Drug and dose/kg body wt	Pre-treatment period	Post-treatment period							% of EPG Reduction
			4th day	8th day	12th day	16th day	20th day	24th day	
		EPG	EPG	EPG	EPG	EPG	EPG	EPG	
A	Control	1000±21.12	1010±17.76	1020±12.36	1020±10.92	1030±9.86	1040±12.36	1060±10.83	
B	Bironja (3ml/kg)	990±29.39	990±20.86	970±10.43	920±19.83	890±11.91	880**±8.23	870**±5.83	13.13%
C	Bironja (6ml/kg)	990±17.74	940±21.32	900±12.21	820±9.32	750**±20.03	610**±10.92	580**±31.32	41.41%
D	Bironja (9ml/kg)	1010±9.89	970±10.56	940±17.07	640**±19.03	500**±11.03	490**±2.56	480**±8.83	51.47%
E	Nilzan (30mg/kg)	980±4.78	620**±22.25	430**±13.94	370**±7.56	250**±8.65	260**±4.53	130**±11.25	86.73%

Values given above represent the mean ± SE of 5 sheep.

**Significant decrease (p<0.01).

Turmeric

In case of Fasciola spp. in group G (2 ml/kg), the mean EPG was 880 and 780 respectively at pre-treatment and 24th day of treatment. In group H, the mean EPG 910 and 400 respectively at pre-treatment and 24th day of treatment following administration of alcoholic extracts of Turmeric at the dose rate 4ml/kg body wt/ sheep. In group I (6 ml/kg) the mean EPG values were 900 and 770, respectively at pre-treatment and 24th day after treatment. Following Nilzan administration (30 mg /kg body wt.) the mean EPG was 900 and 120 respectively at pre-treatment and 24th day of medication .All the three doses of turmeric i.e 2,4 and 6 ml/kg body wt could reduce the mean EPG upto 11-14%. The results were not satisfactory in comparison to Bironja against Fasciola in sheep. However, Nilzan gave similar result as observed in previous study (87% mean EPG reduction).

In case of paramphistomum spp, in group G (2ml/kg), the mean EPG was 820 and 720 respectively at pre-treatment and 24th day of medication. Similarly in group H, the mean EPG was 880 and 760 respectively at pre-treatment and 24th day of medicaton. Following administration of alcoholic extracts of turmeric at the dose rate of 4 ml/kg body wt.. In group I (6ml/kg) the mean EPG values were 850 and 730 respectively before treatment and 24th day of treatment. Following Nilzan administration the mean EPG was 830 and110 respectively before and 24th day of medication. The different doses of Turmeric (2-6 ml/kg body wt) could reduce the mean EPG upto 12-14% against Paramphistomum spp. This result was not satisfactory against Paramphistomum spp. but Nilzan (30 mg/kg body wt.) could reduce the mean EPG up to 87% which was highly satisfactory.

Table 3. The comparative efficacy of Turmeric with that of patent drug Nilzan against *Fasciola spp* in sheep

Group of animals	Drug and dose/kg body wt	Pre-treatment period EPG	Post-treatment period 4th day EPG	8th day EPG	12th day EPG	16th day EPG	20th day EPG	24th day EPG	% of EPG Reduction
F	Control	950±32.46	960±8.96	970±4.56	980±11.58	990±16.32	1000±9.82	1010±10.58	
G	Turmeric(2ml/kg)	880±16.78	870±12.56	860±22.32	840±14.86	800**±7.68	790**±19.35	780**±22.32	11.36%
H	Turmeric(4ml/kg)	910±7.86	890±6.44	870±23.53	850±18.58	830**±6.37	820**±8.56	800**±13.36	12.08%
I	Turmeric(6ml/kg)	900±14.54	880±7.38	860±9.56	840**±4.53	800**±6.95	790**±17.58	770**±8.63	14.44%
J	Nilzan(30mg/kg)	900±32.46	820**±8.96	550**±22.32	420**±4.86	290**±14.35	150**±10.58	120**±8.63	86.66%

Values given above represent the mean ± SE of 5 sheep.

**Significant decrease (p<0.01).

Table 4. The comparative efficacy of Turmeric with that of patent drug Nilzan against *Paramphistomum spp* in sheep.

Group	Drug and dose/kg body wt	Pre-treatment period EPG	Post-treatment period 4th day EPG	8th day EPG	12th day EPG	16th day EPG	20th day EPG	24th day EPG	Percentage of EPG Reduction
F	Control	820±22.67	830±11.11	840±9.03	860±19.86	850±7.03	860±11.92	880±22.03	
G	Turmeric (2ml/kg)	820±14.08	810±8.10	800±6.08	770±22.75	750±23.70	730**±12.95	720**±21.93	12.34%
H	Turmeric (4ml/kg)	880±8.80	860±7.86	840±11.45	820±21.08	790**±9.70	770**±12.76	760**±6.76	12.50%
I	Turmeric (6ml/kg)	850±29.67	840±31.13	820±3.20	790±5.97	770±21.72	750**±22.50	730**±7.96	14.11%
J	Nilzan (30mg/kg)	830±11.32	640**±15.01	480**±19.37	200**±21.11	210**±29.36	150**±31.03	110**±6.89	86.74%

Values given above represent the mean ± SE of 5 sheep.
**Significant decrease (p<0.01).

Veranda

The alcoholic extracts of Veranda leaves in doses of 2, 4, 6 ml/kg body weight orally was almost ineffective against Fascioliasis in sheep. In group L(2 ml/kg) the mean EPG values were 910 and 860, respectively, in group M, the mean EPG values were 980 and 920 respectively(4ml/kg), in group N receiving highest dose (6 ml/kg) the mean EPG values were 960 and 890 respectively at pretreatment and 24th day of treatment. In group O, following administration of Nilzan (30 mg/kg body wt.) the mean EPG values were 930 and 100 respectively at pre-treatment and 24th day after medication. In case of Paramphistomum spp. In group L (2 ml/kg) the mean EPG values were 930 and 860 respectively, in group m, the mean EPG values were 730 and 690, in group N (6 ml/ kg) the mean EPG values were 870 and 800 respectively at pre-treatment and 24th day of treatment. Following administration of Nilzan (30 ml/kg) the mean EPG values were 830 and 100 respectively at pre-treatment and 4th day of medication. The three different doses of alcoholc extracts of veranda leaves i.e 2, 4 and 6 ml/ kg body wt c (the mean EPG reduction upto 4-7% against Fasciola spp. and paramphistomum spp.

Table -5: The comparative efficacy of Verenda with that of patent drug Nilzan against Fasciola spp in sheep.

Group	Drug and dose/kg body wt	Pre-treatment period EPG	Post-treatment period						% of EPG Reduction
			4th day EPG	8th day EPG	12th day EPG	16th day EPG	20th day EPG	24th day EPG	
K	Control	920±19.50	930±17.56	940±13.63	950±2.53	990±11.59	960±2.64	970±6.95	
L	Verenda (2ml/kg)	910±23.52	890±5.86	880±8.43	880±7.35	870±2.36	860±11.09	860±4.46	5.49%
M	Verenda (4ml/kg)	980±15.84	970±21.32	960±14.32	950±11.57	940±18.35	930±16.67	920.65±12.56	6.12%
N	Verenda (6ml/kg)	960±18.36	950±3.52	930±12.54	920±9.86	910±2.76	900±6.87	890.89±7.73	7.29%
O	Nilzan (30mg/kg)	930±14.32	740**±5.69	600**±22.03	430**±2.64	390**±7.35	160**±11.31	100.63**±4.95	88.24%

Values given above represent the mean ± SE of 5 sheep.
**Significant decrease (p<0.01).

Table-6: The comparative efficacy of Verenda with that of patent drug Nilzan against *Paramphistomum spp* in sheep.

Group	Drug and dose/kg body wt	Pre-treatment period EPG	Post-treatment period						% of EPG Reduction
			4th day EPG	8th day EPG	12th day EPG	16th day EPG	20th day EPG	24th day EPG	
K	Control	850±16.86	860±17.42	870±2.32	890±12.53	900±16.32	920±12.86	930±18.23	
L	Verenda (2ml/kg)	930±29.62	920±15.56	910±19.32	900±7.98	890±20.91	880±13.36	880±7.69	5.37%
M	Verenda (4ml/kg)	730±14.68	720±22.35	710±16.78	710±18.36	700±16.96	690±13.56	690±19.56	5.47%
N	Verenda (6ml/kg)	870±18.83	862±6.36	840±22.35	830±16.38	820±17.56	810±13.89	800±19.58	8.04%
O	Nilzan (30mg/kg)	830±6.85	620**±12.36	480**±8.56	320**±6.57	240**±12.35	180**±8.83	100.35**±23.36	87.95%

Values given above represent the mean ± SE of 5 sheep.
**Significant decrease (p<0.01).

Determinatio of biochemical and haematological parameters

Biochemical parameters like serum glucose and SGPT(serum glutamic pyruvic transaminase) were also determined by using autoanalyser (Model No.Reflotron M-06).

Table-7: Effect of alcoholic extracts of Bironja and patent drug Nilzan on SGPT(U/L) in sheep.

Group	Drug and dose/kg body wt P.O.	Pre-treatment period 0	4th day	8th day	12th day	16th day	20th day	24th day
A	Control	10.32±0.02	10.34±0.02	10.42±0.02	10.48±0.03	10.54±0.03	10.58±0.05	10.42±0.07
B	Bironja (3ml/kg)	6.72±0.03	6.73±0.0312	6.74±0.0294	6.82±0.0152	6.76±0.0327	6.78±0.02	6.74±0.01
C	Bironja (6ml/kg)	8.58±0.01	8.57±0.01	8.56±0.0179	8.58±0.0159	8.59±0.0147	8.60±0.01	8.59±0.08
D	Bironja (9ml/kg)	8.01±0.06	7.98±0.03	8.01±0.03	8.02±0.05	8.03±0.06	8.14±0.05	8.26±0.07
E	Nilzan (30mg/kg)	8.12±0.07	7.836±0.06	7.712±0.04	7.814±0.04	7.75±0.04	7.80±0.04	7.82±0.07

Values given above represent the mean ± SE of 5 sheep.

Table -8: Effect of alcoholic extracts of Bironja and patent drug Nilzan on Glucose(mg/dl) in sheep.

Group	Drug and dose/kg body wt P.O.	Pre-treatment period 0	4th day	8th day	12th day	16th day	20th day	24th day
A	Control	85.20±0.04	85.28±0.04	85.42±0.08	85.42±0.02	85.34±0.07	85.42±0.02	85.39±0.41
B	Bironja (3ml/kg)	84.40±0.06	84.45±0.05	84.40±0.05	84.44±0.07	84.58±0.06	84.58±0.08	84.52±0.12
C	Bironja (6ml/kg)	88.24±0.10	88.28±0.06	88.36±0.92	88.40±0.93	88.52±0.01	88.60±0.02	88.44±0.09
D	Bironja (9ml/kg)	79.56±0.09	79.48±0.09	79.48±0.07	79.40±0.08	79.44±0.11	79.44±0.10	79.41±0.23
E	Nilzan (30mg/kg)	86.32±0.09	86.36±0.03	86.38±0.05	86.54±0.05	86.54±0.05	86.64±0.07	86.62±0.09

Values given above represent the mean ± SE of 5 sheep

Tablel 9. Effect of alcoholic extracts of Turmeric and patent drug Nilzan on SGPT(U/L) in sheep.

Group	Drug and dose/kg body wt P.O.	SGPT level on different days(U/L) of treatment and Post-treatment period						
		Pre-treatment period	Post-treatment period					
		0	4th day	8th day	12th day	16th day	20th day	24th day
F	Control	9.26±0.16	9.321±0.08	9.23±0.22	9.25±0.20	9.24±0.11	9.26±0.13	9.23±0.26
G	Turmeric(2 ml/kg)	5.58±0.36	8.55±0.10	8.56±0.21	8.58±0.40	8.56±0.42	8.57±0.26	8.53±0.78
H	Turmeric(4 ml/kg)	9.39±0.12	9.41±0.25	9.46±0.32	9.48±0.19	9.49±0.14	9.45±0.24	9.41±0.25
I	Turmeric(6 ml/kg)	8.76±0.18	8.78±0.26	8.73±0.50	8.76±0.38	8.79±0.09	8.81±0.26	8.83±0.31
J	Nilzan(30m g/kg)	10.25±0.24	10.25±0.10	10.48±0.17	10.36±0.26	10.34±0.16	10.26±0.07	10.35±0.2 2

Values given above represent the mean ± SE of 5 sheep.

Table 10. Effect of alcoholic extracts of Turmeric and patent drug Nilzan on Glucose(mg/dl) in sheep.

Group	Drug and dose/kg body wt P.O.	Glucose level on different days(U/L) of treatment and Post-treatment period						
		Pre-treatment period	Post-treatment period					
		0	4th day	8th day	12th day	16th day	20th day	24th day
F	Control	81.86±0.32	82.02±0.10	82.06±0.29	80.99±0.16	81.32±0.09	81.56±0.26	81.58±0.24
G	Turmeric (2ml/kg)	85.38±0.16	85.38±0.21	85.46±0.17	85.40±0.15	85.42±0.23	85.60±0.12	85.63±0.38
H	Turmeric (4 ml/kg)	84.56±0.06	84.65±0.22	84.92±0.16	84.05±0.15	84.38±0.42	84.63±0.56	84.46±0.17
I	Turmeric (6ml/kg)	84.86±0.21	84.62±0.14	84.83±0.08	84.64±0.19	84.92±0.23	84.32±0.38	84.92±0.10
J	Nilzan (30mg/kg)	83.39±0.43	83.93±0.29	83.43±0.14	83.64±0.24	84.13±0.33	84.64±0.17	84.56±0.18

Values given above represent the mean ± SE of 5 sheep.

Table 11. Effect of alcoholic extracts of Veranda and patent drug Nilzan on SGPT(U/L) in sheep.

Grou p	Drug and dose/kg body wt P.O.	SGPT level on different days(U/L) of treatment and Post-treatment period						
		Pre-treatment period	Post-treatment period					
		0	4th day	8th day	12th day	16th day	20th day	24th day
K	Control	8.85±0.32	8.83±0.31	8.76±0.38	8.81±0.26	8.79±0.09	8.78±0.26	8.75±0.26
L	Verenda (2ml/kg)	8.72±0.51	8.58±0.40	8.25±0.31	8.52±0.77	8.56±0.21	8.62±0.42	8.85±0.35
M	Verenda (4ml/kg)	9.38±0.12	9.31±0.25	9.25±0.20	9.36±0.31	9.48±0.19	9.41±0.25	9.56±0.18
N	Verenda (6ml/kg)	10.11±0.18	10.25±0.24	10.15±0.1	10.26±0.07	10.56±0.1	10.38±0.1	10.36±0.26
O	Nilzan (30mg/kg)	10.23±0.21	10.25±0.10	10.16±0.1	10.46±0.1	10.52±0.1	10.25±0.	10.32±0.2

Values given above represent the mean ± SE of 5 sheep.

Table 12. Effect of alcoholic extracts of Verenda and patent drug Nilzan on Glucose(mg/dl) in sheep

Group	Drug and dose/kg body wt P.O.	Glucose level on different days(U/L) of treatment and Post-treatment period						
		Pre-treatment period	Post-treatment period					
		0	4th day	8th day	12th day	16th day	20th day	24th day
K	Control	88.02±0.32	87.92±0.11	88.12±0.16	87.67±0.09	88.12±0.10	88.18±0.32	87.32±0.21
L	Verenda(2ml/kg)	85.55±0.17	84.95±0.08	85.02±0.12	85.95±0.22	89.32±0.18	84.92±0.19	85.15±0.22
M	Verenda c (4ml/kg)	84.95±0.13	84.42±0.15	84.64±0.26	84.95±0.38	83.98±0.15	84.04±0.53	84.64±0.23
N	Verenda (6ml/kg)	86.32±0.43	85.92±0.12	85.85±0.63	86.12±0.09	86.21±0.16	85.32±0.13	85.12±0.25
O	Nilzan(30mg/kg)	85.55±0.15	85.65±0.56	84.95±0.29	85.85±0.21	84.92±0.18	84.82±0.12	85.05±0.13

Values given above represent the mean ± SE of 5 sheep.

Following treatment with alcoholic extracts of Bironja, Turmeric and Verenda leaves and patent drug Nilzan activities of SGPT and serum glucose level were not significantly changed.Parasitism is one of the most damaging diseases in sheep. It is established that infestations due to trematode undermine the health and productivity of animals such as loss of milk, meat, generation loss and loss of reproductive rate due to animal parasites to the extent of 50 % in Bangladesh. From this point of view a research was conducted to study i) the comparative efficacy of Bironja, Turmeric and Verenda leaves with patent drug Nilzan against trematodiasis in sheep and ii) the effects of Bironja, Turmeric and Verenda leaves and Nilzan on some biochemical parameters in sheep. Fifteen groups of sheep (each consisting of 5 sheep) naturally infected with various trematodes i.e. *Fasciola spp, Paramphistomum spp,*were used to study the anthelmintic efficacy of Bironja, Turmeric and Verenda leaves and patent drug Nilzan.Administration of three different doses of

Bironja(3,6 & 9ml/kg body wt), Turmeric (2,4 & 6 ml/kg body wt) and Verenda leaves(2,4 & 6 ml/kg body wt) showed 12-15% ,11-15% and 4-8% efficacy against trematodes in sheep.Administration of Nilzan(30mg/kg body wt)orally was highly effective (73-89%) against both species of trematodes mentioned above in sheep. Among the medicinal plants , Bironja seeds was found to be most effective followed by Turmeric. Veranda leaves was found to be totally ineffective against tramatodiasis in sheep.No significant change was observed following treatment with Bironja seeds, Turmeric and Verenda leaves and Nizan on serum glucose and SGPT in sheep.

Therefore it can be concluded that the patent drug Nilzan is highly effective against trematodes in sheep. However, among the medicinal plants, Bironja seeds and turmeric may be used against trematodes when patent drugs are not available or become out of reach for the poor farmers of Bangladesh.

REFERENCES

1. Aassan NSBY and Nabi SG, 1987. Comparative efficacy of Banminth-II and Distadin in naturally infected sheep in Kasmir valley. Indian journal in comparative Microbiology, Immunology and Infectious siseases 8: 43-45.
2. Afaz Uddin M, 1985. General Incidence and therapeutic measures of disease in cattle of Savar Military Farm. M. Sc. Thesis submitted to the department of Medicine. B.A.U. Mymensingh.
3. Ananymous (1960): Reported on marketing of wool in Pakistan. Cooperation and marketing Advison, Govt. of Pakistan 1960, 21-22.
4. Arsad MA, Khan AH and Zaman MB, 1956. A note on the plants of medicinal values found in Pakistan. Medicinal plant brance. Pak. Forest Research Institute, Abottabad.
5. Balterwort JH and Morgan ED, 1971. Investigation of the locust feeding inhibition of the seeds of the Neem tree. (Azadirachta indica). Journal of Insect physology, 17: 969-979.
6. Bezubic, B Borowik, M.m and Brzozowska, w./(1978). The effect of panacuron Helminth parasites gnnaturally indected lamb. Actaparasitlogical polanica. 26: 75-82.
7. Boisvencu RJ, Colestock EL and Hendrix JC, 1988. Anthelmintic activity of continuous low doses of fenbendazole into the number of sheep.Veternary parasitology, 26: 321-327.
8. Broughton,H.B,Jones PS,Ley Sv, Morgan ED,Slaurin AMZ, Williams DJ, 1987. The chemical structure of azadirachtis. Proc 3rd Int. Neem cont.Nairobi,Keya 103.
9. Chowdhury II, 1956. Medicinal plants of West Pakistan. Podophylum. Emodil, Pakistan Journal fo Science, 5: 110-116.
10. Faiz MA, 1972. Report on investigation into the epidemiology of parasitic disease in East Pakistan. In Activities of the Research Section of Directorate of livestock services, Bangladesh.1968-1972.
11. Foreyt WJ, 1968. Efficacy of a fenbendazole-Triclabendazole combination against Fasciola hepatica and gastrointestinal nematodes in sheep. Veterinary Parasitology, 26: 265-271.
12. Horak IG, 1978. Parasites of domestic and wild animals in South Africa. V. Helminthes in sheep on dry land pasture on the transval Highland. Onderstepoort Journal of Research.45: 1-5.
13. Joshi HC, 1970. Some pharmacological studies with the seed of Bruten frondosa ornissa. The Veterinary Journal. 5: 5-8.
14. Muktar-Reshid.Mengesha-Fantaye, Feseha-Gebreab, Moges-Moldemeskel, 1993. On the Helminthiasis at Bako. Institute Agricultural Resaerch, Addis Ababa(Ethiopia. Proceedings of 4th national livestock Improvement conference. Addis Ababa(Ethiopia) IAR 1993,P.273-275.
15. Ogunsusi RA, 1978. Changes in blood values of sheep suffering from acute and chronic helminthiasis. Research in veterinary science,25(3):298-301.
16. Soulsby EJL, 1986. Helminth, Arthropodas And protozoa of Domesticated animals, 7th edi,The ELBS and Bailliers, Tindle, Cassell,London,p.216,234,763-766.

11

USE OF DIETARY FENUGREEK (*Trigonella foenum-graecum* L.) SEED FOR THE PRODUCTION OF SAFE BROILER LEAN MEAT

Shah Mohammad Toaha, Bazlur Rahman Mollah and Muslah Uddin Ahammad*

Department of Poultry Science, Faculty of Animal Husbandry,

Bangladesh Agricultural University, Mymensingh-2202, Bangladesh

*Corresponding author: Muslah Uddin Ahammad, E-mail: muslah.ps@bau.edu.bd

RTICLE INFO

ABSTRACT

Key words

Fenugreek seeds,
Broiler,
Productive
performance,
Inclusion level,
Profitability

An experiment was conducted for a period of 28 days to determine the dietary effect of fenugreek seeds (FGS) on the productive and economic performances of broilers. A total of 400 day-old Hubbard Classic straight run broiler chicks were randomly allocated to 5 iso-nitrogenous and iso-caloric dietary treatment groups, each having 80 chicks in 4 replications of 20 numbers, in a completely randomized design (CRD). Broilers were fed *ad libitum* on either basal diet with (positive control) or without (negative control) 0.1% antibiotic (used as antibiotic growth promoter; AGP) or basal diet containing 1.0%, 2.0% or 3.0% FGS. There were no differences in live weight, feed intake and feed conversion ratio (FCR) among the treatment groups for broilers up to 14 days of age (P>0.05). Broilers fed on the diet containing FGS exhibited significantly better productive performances than those fed on AGP (P<0.01). However, the highest productive performances were recorded for the 2.0% FGS-fed broilers in all treatment groups, followed by broilers received 1.0% and 3% FGS (P<0.01). Inclusion of dietary FGS at 2% level resulted in higher dressed carcass, breast, thigh and drumstick meat weight compared to any other level of the FGS inclusion in broiler diet (P<0.01). Unlike the AGP, addition of 2% FGS to the diet significantly reduced abdominal fat (P<0.01). With regard to economic performance, broilers fed on diet containing 2% FGS fetched highest profit in the dietary treatment groups. Inclusion of FGS in broiler diet resulted in lower feed cost and higher profit compared to the inclusion of AGP in the diet. It may be concluded that supplementation of FGS in diets may be useful for efficient and economic production of broiler. The inclusion of FGS at 2% level in broiler diet may be profitable in the production of lean meat of broilers.

INTRODUCTION

It is well recognized that broiler farming is a highly lucrative business enterprise in poultry industry due mostly to the high efficiency of modern broiler strains for converting low quality feed to high quality meat for human consumption as well as their minimal land requirement and quick monetary turnover that in turn encouraging small, medium and large scale poultry farmers to get involved in broiler production. Moreover, the consumption pattern has been shifted from red meat (beef, mutton, lamb, pork) to white meat (broiler meat) due to high saturated fat and cholesterol content of the red meat (Daniel et al., 2011). A regular consumption of red meat has been shown to be directly associated with cardiovascular disease (Micha et al., 2010 and Pan et al., 2012), stroke (Kaluza et al., 2012), type 2 diabetes (Micha et al., 2010 and Pan et al., 2011), obesity (Wang and Beydoun, 2009 and Vergnaud et al., 2010), certain cancers (Cross et al., 2007, Ma and Chapman, 2009 and Pan et al., 2012) and earlier death (Pan et al., 2012). In consequence, the global production and per capita consumption of broiler meat have been increased rapidly in the recent years (Caracalla, 2009).

However, safe broiler meat production always requires maintaining good health, reducing disease outbreak and improving immunity of broilers, because the fast growing broilers are mostly susceptible to invasion of pathogenic microorganisms. Antibiotics are known as health care miracle. They are widely used in veterinary field for reducing the incidence of diseases caused by microorganisms. The routine uses of low-doses or sub-therapeutic-levels of antibiotics often referred to as AGP in broiler feed have been a common practice for more than 50 years to prevent potential diseases as well as to robust gut health, increase meat yield and improve feed efficiency of broilers (Gaskins et al., 2002). However, a large number of studies have provided clear evidence that haphazard use of AGP in broiler feed throughout the production cycle contributes to the accumulation of antibiotic residues in edible meat entering the human food chain, thereby hastening the emergence of antibiotic-resistant bacteria, which poses a dire risk to consumer health (e.g. erosion of the effectiveness of life-saving drugs, persistence of infections and treatment failure) (Nisha, 2008 and Jallailudeen, 2015).

In response to consumer concerns about the safety and ethics of poultry production, the European Union has banned the use of antibiotics in animal production since 2006 (Anonymous, 2005). Subsequently, other developed countries have also limited the antibiotic use in poultry production and most of the feed industries in the developed countries removed all types of antibiotics from poultry feeds and launched the "antibiotic-free" labeled feed (Cogliani, 2011 and Tavernise, 2013). However, the scenario of indiscriminate practice of AGP in poultry feed is still existing in developing countries. It has therefore become a crying need of the time to immediate stop haphazard practicing AGP and start searching for cost-effective and health-promoting alternatives to antibiotics. In the recent years, there has been an increasing trend towards using safe, non-toxic and residue free herbal feed additives (HFA) as potential alternative to AGP. Several studies have shown that inclusion of HFA in broiler diet improves performance, enhances feed utilization and promotes gut health of broiler without having any residual effect on edible meat (Hashemi and Davoodi, 2010). Fenugreek (*Trigonella foenum-graecum* L.) or Methi (Bengali, Hindi and Urdu name) is an annual spicy herbal legume native to Mediterranean regions. It is now cultivated in other parts of the world. It has been reported that FGS contain neurin, biotin, trimethylamine which tends to stimulate the appetite by their action on the nervous system and stimulate growth by increasing the cholesterolemic effects (Al-Jabre, 2003). Chemical analysis studies have revealed that FGS contain protein, fat (especially the linoleic acid), total carbohydrates, vitamins (B-complex, A, D), minerals (calcium, phosphorus, iron, zinc and magnesium), PABA (Para-Amino Benzoic Acid), lecithin, choline, saponins, flavonoids and antibiotic compounds (Michael and Kumawat, 2003; Dixit et al., 2005; Mullaicharam et al., 2013; Mamoun et al., 2014). Studies have shown that it has the properties of lowering blood sugar level, anthelmintic, antibacterial, anti-inflammatory, antipyretic and antioxidant (Xue et al., 2007 and Murlidhar and Goswami, 2012). Some of the beneficial effects of fenugreek in human health include improvement of respiratory, stomach and intestinal health, kidney and liver functions, purification of blood and stimulation of immune system.

In poultry, there have been a few recent studies that tested feasibility of using FGS in the diet for the health benefit of broiler (Ahmad, 2005). Fenugreek seed powder has been shown to decrease plasma total lipids and cholesterol level in commercial broiler (Duru et al., 2013 and Mamoun et al., 2014). Studies also have shown that supplementation of either fenugreek seed extract (Khan et al., 2011), seed powder (Elagib et al., 2013), seed blended with turmeric (Abdel-Rahman et al., 2014) or seed with enzyme (Qureshi et al., 2016) improves the productive performance of broiler chicks. However, no studies have investigated the efficacy of whole fenugreek seed in broiler diet for the production of antibiotic-free lean meat of broiler. Keeping the above facts in view, the current study was therefore designed to examine the effect of different dietary levels of fenugreek seeds on productive performance, carcass characteristics of broiler as well as to assess economic impact of using fenugreek seeds in the diet of broiler.

MATERIALS AND METHODS

This study was conducted with a total of 400 day-old as hatched Hubbard Classic broilers for a period of 28 days to investigate the dietary effect of fenugreek seed on growth performance, feed intake, feed to meat conversion ratio and cost effectiveness of broiler production.

Experimental design
A total of 400 day-old experimental broilers were weighed and randomly distributed to 5 treatment groups (T1 = Basal diet, considered as negative control; T2 = Basal diet supplemented with 0.1% antibiotic (Renamycin), considered as positive control; T3 = Basal diet containing 1.0% fenugreek seed; T4 = Basal diet containing 2.0% fenugreek seed; T5 = Basal diet containing 3.0% fenugreek seed) following a completely randomized design. Each treatment was divided into 4 replicates. Each treatment group had 100 broilers having 20 broilers per replicate. Broilers in each replicate were treated as the experimental unit and diet was treated as the factor.

Formulation of experimental diets
The experimental diets were formulated using locally available quality feed ingredients including fenugreek seed. The formulation of balanced broiler starter and grower diets was done using "user friendly feed formulation done again (UFFDA)" MS Excel program based on the standard nutrient requirement of Hubbard Classic commercial broiler specified in the production manual. After formulation, the 5 starter and 5 grower diets as per treatments were prepared following the standard hand mixing method during the starter (0-14 days) and grower phase (15-28 days), respectively, of broiler rearing. The composition of the ingredients used in the balanced ration formulation and the calculated nutrients profile to meet the nutrient requirement of the broilers are shown in Table 1.

Management of experimental broilers
The broilers were reared in an open-sided, gable type broiler house with concrete floor. The area of the house was divided into 20 equal sized pens partitioned by wire mesh. Rice husk was placed at a depth of 5cm in individual pens. Brooding temperature was maintained at $33^{\circ}C$ for the first week by fixing a 200W-incandescent bulb to each pen. Temperature was then gradually decreased by $2.5^{\circ}C$ each week until the broilers got acclimatized with room temperature. Broilers were exposed to a continuous period of 23 hours lighting and a 1 hour of dark in every 24 hours to encourage full feeding during the period of experiment. Broilers under different feeding regimen were fed on starter diets during the first 14 days, thereafter fed on grower diets until they become 28 days of age. In all cases, feeds were offered *ad libitum* to all broilers. Potable water was made available at all times. Broilers were vaccinated during the experimental period as per recommendation of the manufacturer (Table 2).

Table 1. Composition of broiler starter (0-14 days) and grower (15-28 days) diets and nutrient profiles according to the treatments

Ingredient (%)	Dietary Treatment									
	Negative Control (Basal Diet; BD)		Positive Control (BD with antibiotic)		Fenugreek Seed (%)					
					1		2		3	
	Starter	Grower	Starter	Grower	Starter	Grower	Starter	Grower	Starter	Grower
Maize	60.83	63.83	60.83	63.73	60.22	63.33	59.33	62.33	59.00	61.34
Soybean meal	25.18	22.18	25.10	22.18	25.00	22.00	24.89	21.68	24.52	21.67
Propak	7	6.72	7	6.72	7	6.40	7	6.72	7	6.72
Soybean oil	4	4	4	4	4	4	4	4	3.7	4
Fenugreek seed	0	0	0	0	1	1	2	2	3	3
Dicalcium Phosphate	2.21	2.21	2.19	2.21	2.00	2.21	2.00	2.21	2.00	2.21
Common Salt	0.33	0.33	0.33	0.33	0.33	0.33	0.33	0.33	0.33	0.33
VMA premix*	0.25	0.25	0.25	0.25	0.25	0.25	0.25	0.25	0.25	0.25
DL-Methionine	0.08	0.16	0.08	0.16	0.08	0.16	0.08	0.16	0.08	0.16
L-Lysine	0.12	0.32	0.12	0.32	0.12	0.32	0.12	0.32	0.12	0.32
Renamycin	0	0	0.1	0.1	0	0	0	0	0	0
Total	100.00	100.00	100.00	100.00	100.00	100.00	100.00	100.00	100.00	100.00
Calculated Values of Nutrient Content (%)										
ME (Kcal/kg)	3010	3230	3004	3210	3002	3206	2998	3200	3010	3196
Crude Protein	22.07	20.50	21.93	20.43	22.01	20.32	21.87	20.24	22.07	20.29
Crude Fiber	3.70	3.90	3.70	3.90	3.94	3.98	4.19	4.15	4.25	4.29
Calcium	0.98	0.95	0.95	0.95	0.94	0.95	0.93	0.95	0.93	0.94
Available Phosphorus	0.48	0.48	0.46	0.48	0.45	0.48	0.46	0.48	0.48	0.48
Lysine	1.00	0.97	1.00	0.97	1.00	0.97	1.00	0.97	1.00	0.97
Methionine	0.65	0.65	0.65	0.65	0.65	0.65	0.65	0.65	0.65	0.65
Methionine + Cysteine	0.80	0.80	0.80	0.80	0.80	0.80	0.80	0.80	0.80	0.80
Tryptophan	0.25	0.25	0.25	0.25	0.25	0.25	0.25	0.25	0.25	0.25
Sodium	0.17	0.17	0.17	0.17	0.17	0.17	0.17	0.17	0.17	0.17

*VMA premix, Vitamin-mineral-amino acid premix.

Table 2. Vaccination schedule followed for the experimental broilers

Age (Days)	Name of Disease	Name of Vaccine	Trade Name	Type of Vaccine	Dose	Route of Vaccination
5	ND	BCRDV	BCRDV#	Live	One drop	Ocular
10	IBD	IBD	D-78*	Live	One drop	Ocular
21	IBD	IBD	D-78*	Live	One drop	Ocular

ND, Newcastle Disease; IBD, Infectious Bursal Disease; BCRDV, Baby Chick Ranikhet Disease vaccine, #Directorate of Livestock Services (DLS), Dhaka, Bangladesh; *Intervet International, B.V. BOXMEER, The Netherlands.

Processing of broilers

At the end of feeding trial, one broiler having near to pen average weight was taken from each pen for determining meat yield characteristics. Broilers were slaughtered and allowed to bleed completely. All slaughtered broilers were subjected to semi-scalding (51-55°C) for 120 seconds in order to loose feathers followed by removal of feathers by hand pinning. The procedure of carcass evisceration and dissection was followed as per standard method described by Jones (1984). The viscera, giblet (heart, liver and gizzard) and abdominal fat were removed and weighed for determination of meat yield characteristics. The hot carcass and the individual organs were separately weighted and they expressed as a percentage of live weight. The breast, thigh and drumstick were weighed individually and deboned to yield meat data.

Statistical analysis

All recorded data (body weight, body weight gain, feed intake, FCR, production cost, profitability, dressing yield and dressing percentage) were compiled in Microsoft Excel 2007 and were subjected to analysis of variance (ANOVA) in CRD employing SAS (2009) statistical package program. All significant and non-significant effects were identified by Duncan's New Multiple Range Test (Duncan, 1955).

RESULTS AND DISCUSSION

Live weight

The day-old broiler chicks were weighed individually at the time of random allocation to different dietary treatments. The initial live weights did not differ significantly (P>0.05) between the chicks (Table 3). The differences in body weights of broiler were not significant (P>0.05) during the first and second week of age. However, in relation to age body weights of broilers between the treatment groups tended to differ in response to their different feeding regimen (Table 3). Inclusion of FGS in either broiler starter or grower diet resulted in higher live weight compared to the control diet. Although AGP in the basal diet improved broiler growth, its efficacy was significantly (P<0.05) lower than did the FGS during the last two weeks of age of broiler. Body weight of broilers increased gradually with the increase of FGS inclusion level up to 2% in the diet. Broilers fed on high level (3%) of FGS exhibited significantly (P<0.05) lower growth performance than those on other levels of FGS (Table 3). The highest body weight was observed for the broilers fed on the diet containing 2% FGS during the whole period of feeding trial. Increase of live weight of broilers for the inclusion of FGS in the diet in the current experiment is coincided with the results reported by Abu-Dieyeh and Abu-Darwish (2008), Alloui et al. (2012) and Qureshi et al. (2015). They reported that addition of either whole, crushed or powder form of FGS at various levels in broiler diets increased live weight of broilers. However, the result of the present study contradicts with the findings of Abbas and Ahmed (2010), who reported that broilers supplemented with diet containing 1% or 2% FGS showed low live weight gain. The improvement of live weight of the broilers by FGS might be due to the fact that the FGS contain essential fatty acids and high quality proteins (Murray et al., 1991) as well as have stimulating effect on the villus height of digestive system of broilers (Hernandez et al., 2004; Hind et al., 2013; Mamoun et al., 2014; Mahmood et al., 2015).

Feed intake

The intake of feed by the broilers was increased gradually with the advancement of their age. It is revealed that the feed intake values during the first two weeks were merely numerically different but not statistically significant (P>0.05; Table 3) for the broilers in different dietary groups. Broilers received diet containing 2% FGS consumed the highest amount of feed at all ages followed by the broilers received 1% FGS, 3% FGS, AGP in the diet and basal diet alone. In this study it has been found that FGS when added in the diet up to the level of 2% stimulated feed intake of broilers which are in agreement with the findings of Michael and Kumawat (2003) and Alloui et al. (2012). It has also been shown that FGS contain galactomannans, neurin, biotin, trimethylamine which tends to stimulate the appetite, improve palatability and digestive process by their action on the nervous system and gut microflora which could be attributed to the increase in feed consumption (Michael and Kumawat, 2003 and Alloui et al., 2012). However, the affected feed consumption by the broilers in 3.0% FGS group in the current study might be due to the presence of bitter taste and pungent odor in FGS (El-Kloub, 2006). Other studies have also shown that addition of high level of ground FGS to the broiler diet decreased feed intake (Durrani et al., 2007 and Abbas and Ahmed, 2010).

Table 3. Live weight, feed intake, feed conversion ratio of broilers fed on different treatments

| Variable (g/broiler) | Age (days) | Dietary Treatment | | | | | Level of significance (LSD[+]) |
| | | Negative Control[A] | Positive Control[B] | Fenugreek seed (%) | | | |
				1	2	3	
Live weight	0	42.9	42.3	42.7	43.1	42.2	NS (1.3)
	7	148.8	155.7	159.5	170.2	156.8	NS (24.7)
	14	350.2	368.8	392.9	402.2	371.9	NS (38.3)
	21	782.2[a]	815.5[a]	852.9[b]	897.2[bc]	822.5[ab]	* (69.5)
	28	999.8[a]	1065.9[a]	1172.4[b]	1251.6[c]	1101.0[gd]	* (68.9)
Feed intake	7	125.3	126.4	128.3	134.7	127.2	NS (8.1)
	14	409.9	428.6	430.9	453.0	429.9	NS (56.8)
	21	1236.4[a]	1257.8[ab]	1281.3[b]	1316.3[c]	1258.8[ab]	* (28.7)
	28	2067.3[a]	2084.2[ab]	2115.9[b]	2158.2[c]	2092.4[ab]	* (45.7)
Feed conversion ratio	7	1.19	1.12	1.10	1.06	1.11	NS (0.1)
	14	1.34	1.33	1.32	1.26	1.31	NS (0.2)
	21	1.67[a]	1.64[ab]	1.58[b]	1.54[bc]	1.62[b]	* (0.04)
	28	2.16[a]	2.04[ab]	1.88[b]	1.79[bc]	1.98[b]	* (0.1)

[A] Negative control, Basal diet; [B] Positive control, Basal diet containing 0.1% Renamycin. [abcd] mean values with dissimilar superscripts are significantly different (P<0.05). [+]LSD, data in the parenthesis indicate the least significant difference value. NS, non-significant, P>0.05; *, P<0.05.

Feed Conversion Ratio

Inclusion of FGS in broiler diet resulted in lower FCR compared to the AGP inclusion in feed throughout the experiment. Broilers fed on basal diet alone exhibited the highest FCR at all ages. The differences in FCR were significant (P<0.05; Table 3) between the dietary treatment groups following the second week of age of the broilers. High inclusion (3%) level of FGS in the diet gave rise to higher FCR than any other inclusion level of FGS (Table 3). The present study revealed that FCR was lowered by the feeding FGS to the broilers compared to AGP feeding to broilers which is coincided with the study by Abu-Dieyeh and Abu-Darwish (2008) and AL-Beitawi and El-Ghousein (2008). Improvement in feed efficiency by FGS might be related to the development of the broiler chicks gut morphological changes of gastrointestinal tissues (Alloui et al., 2012; Amal et al., 2013; Mukhtar et al., 2013; Weerasingha and Atapattu, 2013)

Meat yield characteristics

Significant differences were obtained for the percentage of dressed weight, breast meat and abdominal fat between the treatment groups (P<0.05; Table 4). Broilers fed on diet containing 2% FGS yielded the highest meat in the breast and the lowest fat in the abdomen (Table 4). Inclusion of AGP in the diet however resulted in higher deposition of abdominal fat in broilers. Feeding control diet (basal diet alone) resulted in increased liver weight of broilers. There were no significant differences among treatment groups for the non-carcass (heart, liver and gizzard) yields (P>0.05; Table 4). It has been appeared from the present study that the weights of dressed carcass, thigh, drumstick and breast of broilers were increased and the weights of gizzard, liver and heart were decreased due to the inclusion of FGS in the diet. These results are in line with the findings of Durrani et al. (2007), Abbas and Ahmed (2010) and Alloui et al. (2012) who showed that broilers fed on 1% and 2% fenugreek seed powder exhibited higher dressing percentages compared to unsupplemented group. The current study results are in agreement with the reports of Mukhtar et al. (2013), who showed that feeding fenugreek seed powder to broilers decreased non carcass component (liver, gizzard and heart) weight. However, El-Ghammry et al. (2002) found that broiler chicks fed on diet containing very low levels (0.2% and 0.4%) of crushed FGS had low dressing percentage. However, in contradiction with the results of the current research Al-Beitawi and El-Ghousein (2008) reported that the supplementation of different levels of crushed or uncrushed fenugreek seed did not affect any of the carcass characteristics parameters. However, feeding FGS at 2% level clearly reduced the abdominal fat content in the present study. The abdominal fat is harmful for human health. Studies have shown that the FGS have the fat reducing activity by the presence of lecithin and choline in FGS (Dixit et al., 2005).

Table 4. Meat yield characteristics of broilers fed on different treatments

Variable (%)	Dietary Treatment					Level of significance LSD[+]
	Negative Control[A]	Positive Control[B]	Fenugreek seed (%)			
			1	2	3	
Dressed weight	54.33[a]	55.6[a]	61.10[b]	67.20[c]	57.10[a]	*(2.80)
Breast meat	11.54[a]	12.63[b]	13.77[c]	15.14[d]	2.34[b]	*(0.52)
Thigh meat	7.90	8.78	9.23	10.15	8.64	NS (1.63)
Drumstick meat	5.31	5.56	5.64	6.24	5.55	NS (0.80)
Abdominal fat	1.24[a]	1.48[b]	1.12[a]	0.94[c]	1.26[a]	*(0.15)
Gizzard	2.06	1.84	1.77	1.73	2.02	NS (1.04)
Heart	0.56	0.55	0.51	0.52	0.50	NS (0.06)
Liver	2.79	2.49	2.43	2.41	2.36	NS (0.57)

[A] Negative control, Basal diet; [B] Positive control, Basal diet containing 0.1% Renamycin. [abcd] mean values with dissimilar superscripts are significantly different (P<0.05). [+]LSD, data in the parenthesis indicate the least significant difference value. NS, non-significant, P>0.05; *, P<0.05.

Table 5. Cost of production and profit of broilers fed on different treatments

Variable	Dietary Treatment					Level of significance (LSD+)
	Negative Control[A]	Positive Control[B]	Fenugreek seed (%)			
			1	2	3	
Final body weight (kg/broiler)	999.8[a]	1065.9[a]	1172.4[b]	1251.6[c]	1101.0[ad]	*(68.9)
Total feed intake (kg/broiler)	2067.3[a]	2084.2[ab]	2115.9[b]	2158.2[c]	2092.4[ab]	*(45.7)
Feed price (BDT/kg)	33.05	33.62	33.20	33.60	33.80	-
Feed cost (BDT/kg broiler)	68.37[a]	65.80[a]	60.01[b]	57.99[b]	64.42[a]	*(4.47)
Feed cost (BDT/broiler)	68.32[a]	70.07[b]	70.25[b]	72.51[c]	70.72[b]	*(1.52)
Other costs (BDT/broiler)	40.00	40.00	40.00	40.00	40.00	-
Total production cost (BDT/broiler)	108.32[a]	110.07[b]	110.25[b]	112.51[c]	110.72[b]	*(1.51)
Total production cost (BDT/kg broiler)	108.37[a]	105.80[a]	100.01[bd]	97.99[cd]	104.42[ab]	*(4.47)
Return/broiler (Sale price BDT120.00/kg broiler)	119.98[a]	127.91[a]	140.69[b]	150.19[c]	132.12[a]	*(8.27)
Profit (BDT/broiler)	11.65[a]	17.84[ab]	30.44[c]	37.67[c]	21.39[b]	*(8.92)
Profit (BDT/kg broiler)	11.61[a]	16.64[ab]	25.82[c]	30.02[c]	19.13[b]	*(6.57)
Profit (BDT/kg broiler) over the control	-	5.03[a]	14.21[bc]	18.41[cd]	7.52[ab]	*(7.13)

[A] Negative control, Basal diet; [B] Positive control, Basal diet containing 0.1% Renamycin. [abcd] mean values with dissimilar superscripts are significantly different (P<0.05). [+]LSD, data in the parenthesis indicate the least significant difference value. NS, non-significant, P>0.05; *, P<0.05. BDT, Currency in Bangladesh. Conversion: 1 BDT = 0.013 USD or 0.012 EUR or 0.010 GBP.

Cost-effectiveness of broiler production

The calculated per kg feed prices for the basal diet alone, basal diet containing AGP, 1% FGS, 2% FGS and 3% FGS were BDT 33.05, BDT 33.62, BDT 33.20, BDT 33.60 and BDT 33.80, respectively. The price of feed per kg was increased for the inclusion of either AGP or FGS in the basal diet. However, higher feed price was recorded for the high inclusion level (2% and 3%) of FGS compared to either AGP or low level (1%) of FGS inclusion in the basal diet. The calculated total production costs per kg broiler were very high, high, moderate, low and very low for the basal diet alone, basal diet containing AGP, 3% FGS, 1% FGS and 2% FGS, respectively (P<0.05; Table 5). The net income in terms of per broiler generated from the sale of live broiler was the highest for 2% FGS treatment group followed by 1% FGS, 3% FGS, AGP and negative control group (Table 5). Consequently, the net profit per kg broiler was also recorded highest for the broilers fed on

2% FGS followed by 1% FGS, 3% FGS, AGP and basal diet alone (P<0.05; Table 5). Inclusion of FGS in broiler diet resulted in lower feed cost and higher profit compared to the inclusion of AGP in the diet. The present study clearly indicates that feeding FGS was the most cost-effective and had beneficial effect on profitability of broiler. These results were in line with the findings of Mukhtar et al. (2013), who reported that supplementation of fenugreek seed powder to broiler diet resulted in economic benefits. Mamoun et al. (2014) also recorded profit for the broilers fed on diets containing 1% FGS compared to 3% FGS group. Increase in the profitability of broilers fed rations containing herbal growth promoters may be attributed to the better efficiency of feed utilization, which resulted in more growth and better conversion feed to live weight gain.

CONCLUSIONS

On the whole, it is therefore concluded that the 2% level of FGS was superior in terms of live weight, FCR and cost-effectiveness over AGP and other levels of FGS in the diet. The 2% level of FGS can be used in broiler diets as an alternative to AGP for economic and efficient production of lean broilers.

REFERENCES

1. Abbas TE and ME Ahmed, 2010. Fenugreek for poultry feed. World's Poultry Science Journal, 23: 66-80.
2. Abdel-Rahman HA, SI Fathallah, AA Helal, AA Nafeaa and IS Zahran, 2014. Effect of turmeric (*Curcuma longa*), fenugreek (*Trigonella foenum-graecum* L.) and/or bioflavonoid supplementation to the broiler chick's diet and drinking water on the growth performance and intestinal morphometric parameters. Global Veterinaria, 12: 627-635.
3. Abu-Dieyeh ZHM and MS Abu-Darwish, 2008. Effect of feeding powdered fenugreek seeds *(Trigonella foenum L.)* on growth performance of 4-8 week-old broilers. Journal of Animal and Veterinary Advances, 7: 286-290.
4. Ahmad S, 2005. Comparative efficiency of garlic, turmeric and kalongi as growth promoter in broiler. M.Sc. Thesis, Department Poultry Sciences, University of Agriculture, Faisalabad, Pakistan.
5. Al-Betawi N and SS El-Ghousein, 2008. Effect of feeding different levels of *Trigonella foenum* seeds (Fenugreek) on performance, blood constituents, and carcass characteristics of broiler chicks. International Journal of Poultry Science, 7: 715-721.
6. Al-Jabre S, Al-Akloby, OM Qurashi, AR Akhtar, A Al-Dossary and MA Randhawa, 2003. Thymoquinone, an active principle of *Trigonella foenum*, inhibited *Aspergillus niger*. Pakistan Journal of Medical Research, 42: 102-104.
7. Alloui N, SB Aksa and MN Alloui, 2012. Utilization of fenugreek (*Trigonella foenum graecum*) as growth promoter for broiler chickens. Journal of World's Poultry Research, 2: 25-27.
8. Amal O, AMA Mukhtar, KA Mohamed and H Ahlam, 2013. Use of half bar essential oil (HBO) as a natural growth promoter in broiler nutrition. International Journal of Poultry Science, 12: 15-18.
9. Anonymous, 2005. European Commission ban on antibiotics as growth promoters in animal feed enters into effect. Available at: http://europa.eu/rapid/press-release_IP-05-1687_en.htm.
10. Caracalla VT, 2009. The state of food and agriculture. Food and Agriculture Organization of the United Nations. FAO, 00153 Rome, Italy, pp180.
11. Cogliani C, H Goossens and C Greko, 2011. Restricting antimicrobial use in food animals: lessons from Europe banning nonessential antibiotic uses in food animals is intended to reduce pools of resistance genes. Microbe, 6: 274-279.
12. Cross AJ, MF Leitzmann, MH Gail, AR Hollenbeck, A Schatzkin and R Sinha, 2007. A prospective study of red and processed meat intake in relation to cancer risk. PLOS Medicine, 4: 1974-1984.
13. Daniel CR, AJ Cross, C Koebnick and R Sinha, 2011. Trends in meat consumption in the USA. Public Health Nutrition, 14: 575–583.
14. Dixit P, S Ghaskadbi, H Mohan and TPA Devasagayam, 2005. Antioxidant properties of germinated fenugreek seeds. Phytotherapy Research, 19: 977–983.
15. Duncan, DB, 1955. Multiple range and multiple 'F' test. Biometrics, 11: 1-42.

16. Durrani FR, N Chand, K Zaka, A Sultan, FM Khattak and Z Durrani, 2007. Effect of different levels of feed added fenugreek on the performance of broiler chicks. Pakistan Journal of Biological Science, 10: 4164-4167.

17. Duru M, Z Erdoğan, A Duru, A Küçükgül, V Düzgüner, D Alpaslan and A Şahin, 2013. Effect of seed powder of herbal legume fenugreek (*Trigonella foenum-graceum* L.) on growth performance, body components, digestive parts and blood parameters of broiler chicks. Pakistan Journal of Zoology, 45: 1007-1014.

18. Elagib HAA, SA Abbas and KM Elamin, 2013. Effect of different natural feed additives compared to antibiotic on performance of broiler chicks under high temperature. Bulletin of Environment, Pharmacology and Life Sciences, 2: 139-144.

19. El-Ghammry AA, GM El-Mallah and AT El-Yamny, 2002. The effect of incorporation yeast culture, *Trigonella foenum* seeds and fresh garlic in broiler diets on their performance. Egyptian Journal of Poultry Science, 22: 445-459.

20. El-Kloub M, 2006. Effect of using commercial and natural growth promoters on the performance of commercial laying hens. Egyptian Poultry Science, 26: 941-965.

21. Gaskins HR, CT Collier and DB Anderson, 2002. Antibiotics as growth promotants: mode of action. Animal Biotechnology, 13: 29-42.

22. Hashemi SR and H Davoodi, 2010. Phytogenics as new class of feed additive in poultry industry. Journal of Animal and Veterinary Advances, 9: 2295-2304.

23. Hernandez AI, J Madrid and V Garcia, 2004. Influence of two plant extracts on broiler performances, digestibility and digestive organs size. International Journal of Poultry Science, 83: 169-174.

24. Hind AA, AA Saadia and ME Khalid, 2013. Effect of different natural feed additives compared to antibiotic on performance of broiler chicks under high temperature environment. Bulletin of Environment, Pharmacology and Life Sciences, 2: 139-144.

25. Jallailudeen RL, MJ Saleh, AG Yaqub, MB Amina, W Yakaka and M Muhammad, 2015. Antibiotic residues in edible poultry tissues and products in Nigeria: A potential public health hazard. International Journal of Animal and Veterinary Advances, 7: 55-61.

26. Jones, 1984. A standard method for the dissection of poultry for carcass analysis. The West of Scottland Agricultural College, Auchincruive, Ayr. Technical Note, 222.

27. Kaluza J, A Wolk and SC Larsson, 2012. Red meat consumption and risk of stroke: a meta- analysis of prospective studies. Stroke, 43: 2556-2560.

28. Khan FU, A Assad, SU Rehman, S Naz and N Rana, 2011. Fenugreek (*Trigonella foenum-graecum* L.) effect on muscle growth of broiler chicks. Research Opinions in Animal and Veterinary Sciences, 1: 1-3.

29. Ma RW and K Chapman, 2009. A systematic review of the effect of diet in prostate cancer prevention and treatment. Journal of Human Nutrition and Diet, 22: 187-99.

30. Mahmood S, A Rehman, M Yousaf, P Akhtar, G Abbas, K Hayat, A Mahmood and MK Shahzad, 2015. Comparative efficacy of different herbal plant's leaf extract on haematology, intestinal histomorphology and nutrient digestibility in broilers. Advances in Zoology and Botany, 3: 11-16.

31. Mamoun T, MA Mukhtar and MH Tabid, 2014. Effect of fenugreek seed powder on the performance, carcass characteristics and some blood serum attributes. Advance Research in Agriculture and Veterinary Science, 1: 6-11.

32. Micha R, SK Wallace and D Mozaffarian, 2010. Red and processed meat consumption and risk of incident coronary heart disease, stroke, and diabetes mellitus: a systematic review and meta-analysis. Circulation, 121: 2271-2283.

33. Michael D and D Kumawat, 2003. Legend and archeology of fenugreek, constitutions and modern applications of fenugreek seeds. International Sympozium, USA, pp 41-42.

34. Mukhtar MA, KA Mohamed, OA Amal and H Ahlam, 2013. Response of broiler chicks to different dietary spearmint oil as a natural growth promoter. University of Bakht Alruda Scientific Journal, 6: 175-183.

35. Murlidhar M and TK Goswami, 2012. A review on the functional properties, nutritional content, medicinal utilization and potential application of fenugreek. Journal of Food Processing and Technology, 3: 1-10.

36. Murray RK, DK Granner, PA Mayes and VW Rodwell, 1991. The text book of Harpers Biochemistry. 22nd Edition. Applecton and large. Norwalk, Connecticut/Loss Altos, California.

37. Nisha AR, 2008. Antibiotic residues – A global health hazard. Veterinary World, 1: 375-377.

38. Pan A, Q Sun , AM Bernstein, MB Schulze, JE Manson, WC Willett and FB Hu, 2011. Red meat consumption and risk of type 2 diabetes: 3 cohorts of US adults and an updated meta-analysis. American Journal of Clinical Nutrition, 94: 1088-1096.

39. Pan A, Q Sun, AM Bernstein, MB Schulze, JE Manson, MJ Stampfer, WC Willett, FB Hu, 2012. Red meat consumption and mortality: results from 2 prospective cohort studies. Archives of Internal Medicine, 172: 555-563.

40. Qureshi S, MT Banday, I Shakeel and S Adil, 2016. Feeding value of raw or enzyme treated dandelion leaves and fenugreek seeds alone or in combination in meat type chicken. Pakistan Journal of Nutrition, 15: 9-14.

41. Qureshi S, MT Banday, S Adil, I Shakeel and ZH Munshi, 2015. Effect of dandelion leaves and fenugreek seeds with or without enzyme addition on performance and blood biochemistry of broiler chicken and evaluation of their *in vitro* antibacterial activity Indian Journal of Animal Sciences, 85: 1248-1254.

42. Tavernise S, 2013. The Food and Drug Administration restricts antibiotics use for livestock. The New York Times (Health section), New York, USA. December 11, 2013.

43. Vergnaud AC, T Norat, D Romaguera, T Mouw, AM May, N Travier, J Luan, N Wareham, N Slimani and S Rinaldi *et al.*, 2010. Meat consumption and prospective weight change in participants of the EPIC-PANACEA study. American Journal of Clinical Nutrition, 92: 398-407.

44. Wang Y and MA Beydoun, 2009. Meat consumption is associated with obesity and central obesity among US adults. International Journal of Obesity, 33: 621-628.

45. Weerasingha AS and NSBM Atapattu, 2013. Effects of Fenugreek (*Trigonella foenum-graecum* L.) seed powder on growth performance, visceral organ weight, serum cholesterol levels and the nitrogen retention of broiler chicken. Tropical Agricultural Research, 24: 289-295.

46. Xue WL, XS Li, J Zhang, YH Liu, ZL Wang and RJ Zhang, 2007. Effect of *Trigonella foenum-graecum* (fenugreek) extract on blood glucose, blood lipid and hemorheological properties in streptozotocin induced diabetic rats. Asia Pacific Journal of Clinical Nutrition, 16: 422-426.

12

EFFICACY OF NEEM LEAF, NISHYINDA LEAF AND TURMERIC RHIZOME SUPPLEMENTATION ON THE GROWTH PERFORMANCE OF BROILER CHICKEN

Nahid Nawrin Sultana[1]*, Soheli Jahan Mou[2], Mahbub Mostofa[1] and Md. Abdur Rahman[3]

[1]Department of Pharmacology and [2]Department of Surgery and Obstetrics, Faculty of Veterinary Science, Bangladesh Agricultural University (BAU), Mymensingh-2202, Bangladesh; [3]Concultant Veterinarian, Veterinary Teaching Hospital, BAU, Mymensingh-2202, Bangladesh.
Present affiliation: [1]Veterinary Surgeon, Upazilla Livestock Office, Haluaghat, Mymensingh; [2]Veterinary Surgeon, Upazilla Livestock Office, Islampur, Jamalpur

*Corresponding author: Nahid Nawrin Sultana; E-mail: nahidkanta01@gmail.com

RTICLE INFO

ABSTRACT

Key words

Efficacy,
Medicinal plants,
Haematology,
Growth,
Broiler

This experiment was conducted to evaluate the efficacy of Neem (*Azadirachta indica*) leaf, Nishyinda (*Vitex nogundo*) leaf and Turmeric rhizome (*Curcuma longa*) powdered supplementation in drinking water as a growth promoter in broiler chickens. A total of 40 day-old Cobb 40 broiler chicks were purchased from local hatchery (Nourish Poultry & Hatchery Ltd.) and after seven days of acclimatization chicks were randomly divided into two groups, A and B. The group A was kept as a control and not treated. The group B was supplemented with Neem, Nishyinda leaves and Turmeric dried powder with feed and water. Weekly observations were recorded for live body weight gain up to 5th weeks and hematological tests were performed at 7th and 35th day's age of broiler to search for hematological change between control (A) and treatment (B) groups. The initial body weight of groups A and B on 7th day of this experiment were 130±4.35 gm, respectively and after 35th day of experiment final body weight were 150±47.35 gm and 1600±58.56 gm, respectively; the net body weight gain were 1320±43.79 gm and 1470±54.25 gm, respectively and economics of production were analyzed and found that net profit per broiler was Tk. 17.24 and Tk. 30.00, respectively. The treatment group B was recorded statistically significant (at 1% level) increase for live body weight than that of control group A. The hematological difference, while Hb. estimation does not show significant difference from control group. The results suggest that better growth performance could be achieved in broilers supplemented with Neem, Nishyinda leaves and Turmeric rhizome extract.

INTRODUCTION

Medicinal plants compete with synthetic drugs, and the majority have no residual effects (Tipu et al., 2006). Emerging health hazards are evident in animals and man by irrational use of antibiotics and antimicrobial growth promoters. As the world is becoming more advanced, new diseases are emerging in animals and human beings by irrational use of antibiotics and antimicrobial growth promoters. Now it is the need of the hour to work more extensively on the medicinal plants in the greater interest of mankind. Antibiotics and inorganic growth promoters are used in the poultry feed to protect the birds from different diseases, to promote growth of the birds, to improve feed conversion ratio (FCR), to increase weight gain and to maximize economic returns from the individual bird. However, these are also being misused. The antibiotic abuse occurs when these are. used unnecessarily, over prescribed, employed in wrong combination, changed quickly over to the other drugs, used persistently, given in inadequate dosage, given in self-medication, used for preventive purposes and employed as unauthorized.

There are many factors leading to the occurrence of antibiotic residues in animal products e.g. failure to observe drug withdrawal period, extended usage or excessive dosages of antibiotics, non-existence of restrictive legislation or their inadequate enforcement, poor records of treatment, lack of advice on withdrawal periods, off-label use of antibiotics, availability of antibiotics to lay persons as over-the counter drugs in the developing countries, lack of consumer awareness about the magnitude and human health hazards associated with antibiotic residues in the food of animal origin. Due to the outbreaks of resistant bacteria and residues of antibiotics in products, currently there are several kinds of antibiotics alternative developed and used. Among the alternatives, medicinal plants with excellent pharmacological activity are getting attention by researchers.

Herbal agents could serve as safer alternatives as growth promoters due to lower cost, reduced toxicity and minimum health hazards. Biological trails of certain herbal formulations as growth promoter have shown encouraging results and some of the reports have demonstrated improved weight gain and feed efficiency, lowered mortality and increased immunity and viability in poultry (Kumar, 1991). Some herbal growth promoters exert therapeutic effects against liver damage due to feed contaminants like aflatoxin (Ghosh, 1992). Bangladesh is abundant in plants possessing interesting pharmacological properties, which await exploitation. Various herbal products are being used as growth promoters in poultry rations like nishyinda, black pepper and cinnamon. Antibiotics promote growth because of an effect on gut flora (de Man, 1975). The use of antibiotics as dietary growth promoters in poultry diets has reduced dramatically. Antimicrobial resistance in zoonotic pathogens including Salmonella spp., E. coli and Enterococci in food animals is of special concern to human health because these are likely to transfer to humans (Endtz et al., 1991). In 2006, the European commission banned the last fourfeed antibiotics (monensin sodium, salinomycin sodium, avilamycin and flavor phospholipol). To minimize resistance, different agencies are in favour of banning these feed antibiotics (Hileman, 2002). The phasing out of antibiotic growth promoters will affect the poultry industry. There is a need to find alternatives. There are a number of alternatives such as enzymes, inorganic acids, probiotics, herbs immunostimulants and management practices (Banerjee, 1998). Herbs and their essential oils have long been known for their antimicrobial activity (Chang, 1995). Polyherbal extracts have been worldwide for a range of medicinal properties like antibacterial, antiviral, antifungal, antiprotozoal, or hepato-protective without adverse effects (Kale et al., 2003; Chowdhury et al., 2009). Nishyinda (*Vitex negundo* L.) is a hardy plant, flouring mainly in the Indian subcontinent. It possesses phyto-chemical secondary metabolites, which impart a variety of medicinal uses. The leaves of nishyinda may be applied locally to swellings from rheumatoid arthritis and sprains. The juice of the leaves is used for the treatment of foetid discharges. The principal constituents of the leaf juice are casticin, isoorientin, chrysophenol D, luteolin, p-hydroxybenzoic acid and D-fructose. Black pepper (Piper nigrum) is a flowering vine in the family Piperaceae, cultivated for its fruit and used as a spice and seasoning. Dried gound pepper has been used for both its flavor and as a medicine, which is due to presence of piperine. Cinnamon (*Cinnamomum zylenicum*) is commonly used in the food industry. It has strong antibacterial, anti-candida, anti-ulcer, analgesic, antioxidant and hypocholesterolaemic activities (Mastura et al., 1999; Lin et al., 2003).

In view of these, the present work had been undertaken to investigate the growth performance of broilers supplemented with Neem leaf, Nishyinda and Turmeric rhizome and to examine the effect of Neem, Nishyinda and Turmeric powder on haematological parameters (TEC, Hb., ESR and PCV) of broilers.

MATERIALS AND METHODS

Broiler chicks and experimental design

The experiment was conducted in the Department of Pharmacology, Faculty of Veterinary Science, Bangladesh Agricultural University, Mymensingh. A total of forty broiler chicks were collected from Nourish Hatchery and allowed to acclimatize for seven days. The birds were kept on a floor litter system in separate pens each measuring 0.9 x 1.2 metres. The pens were thoroughly cleaned, white-washed and disinfected before use. All the birds were provided same management. Fresh clean water was made available at all times.

The birds were randomly divided into two equal groups. Group A was kept as control without any supplement while group B was supplemented with polyherbal extract 1 mL/litre in drinking water. B-groups were received Neem leaves, Nishyinda leaves and Turmeric rhizome powder were used as growth promoter as treatment along with basal diet. The leaves of Neem and Nishyinda and Turmeric rhizome powder were prepared according to procedure of Molla et al. (2012). All the chicks of treated and control groups were closely observed for 35 days and clinical signs were recorded.

Hematological Parameters

Blood samples were collected from wing vein of chicken of both control and treated groups at 7th and 35th days to study the effect Neem, Nishyinda leaves and Turmeric rhizome and the following parameters were observed in the laboratory using standard procedure:

   a. Total Erythrocyte Count (TEC)
   b. Hemoglobin estimation (Hb)
   c. Packed Cell Volume (PCV) and
   d. Erythrocyte Sedimentation Rate (ESR)

Performance trial

During the 35 days experimental period, growth performance was evaluated. Body weight and feed consumption were recorded weekly and body gain and feed conversion were than calculated. Mortality was recorded throughout the study.

Statistical analysis

The data were analysed statistically between control and treated groups of chicken by Student's t-test.

RESULTS

The experimental units were kept on a floor litter system in separate pens. From the present study, table 1 revealed that, in control group (Group A) initial average live weight on 7th day was 130±3.56 gm, final live weight 1450± 47.35 gm, weight gain 1320±43.79 gm and feed conversion ratio (FCR) was 2.20. In Group B initial average live weight on 7th day was 130±4.35 gm, final live weight 1600±58.56 gm, weight gain 1470±54.25 gm and FCR 1.93.

The birds of group B using drinking water were supplemented with 1% Neem, Nishyinda and Turmeric powder utilized their feed statistically significantly (at 1% level) more efficiently than control group (Table 2). Statistical analysis of the data shows non-significant between the dressing percentages of the birds of two groups (Table 2).

Statistical analysis of the data did not show any difference between the relative gizzard weights of the birds of two groups (Table 2). Statistical analysis of the data shows 1% level of significance of relative heart, liver, spleen and pancreas weight between the birds of two groups using drinking water with or without supplementation of Neem, Nishyinda and Turmeric (Table 2).

Economies of Production

The average rearing cost of broilers in two groups were Tk. 179.00 and Tk. 176.00 for A and B groups, respectively (Table 3), excluding the cost of labour because the experiment was conducted on the Department of Pharmacology research shed, Bangladesh Agricultural University, Mymensingh. Miscellaneous cost summed up Tk. 20 per broiler, which included the estimated cost of electricity, litter and disinfectant. The average live weight/broiler in groups A and B were 1.450 kg and 1.600 kg respectively. The broilers were sold in live weight basis at the rate of Tk. 140/kg. The net profit/ Kg live weight in the respective group excluding the cost of labour was found to be Tk.17.24 and Tk. 30.00, respectively.

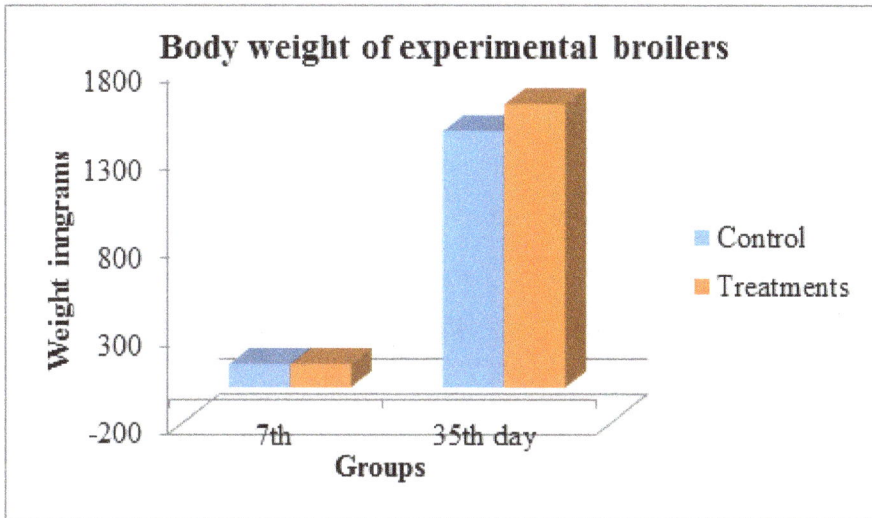

Figure 1. Body weight of experimental chickens

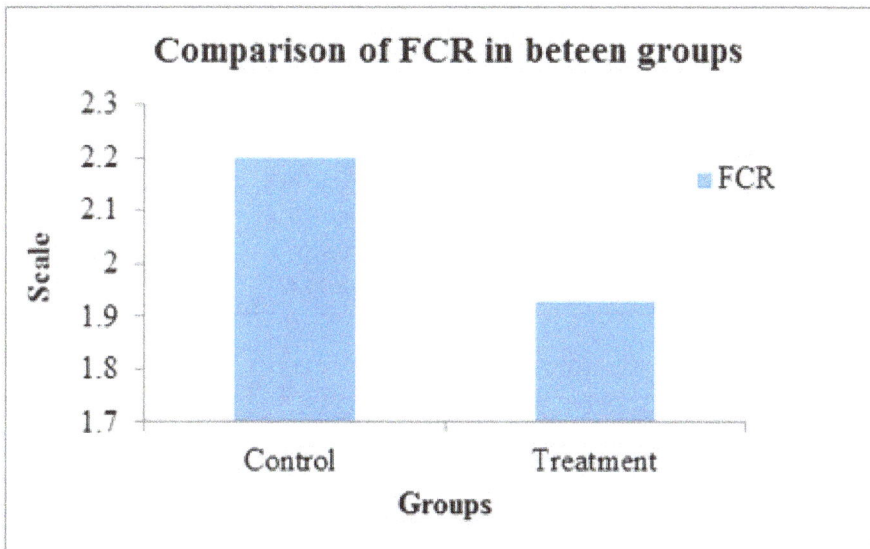

Figure 2. Comparison of feed conversion ration

Table 1. Initial and final live weight, weight gain, feed consumption and feed conversion ratio of broilers supplemented with or without neem leaf, nishyinda leaf and turmeric rhizome on 1-5 weeks of age

Variables	Treatments	Average weight (Mean ± SEM)	P value	Significance level
Initial live weight (g) on 7th day	Control	130±3.56	0.000	**
	Neem, Nishyinda and turmeric	130.4.35		**
Final live weight (g) on 35th day	Control	1450±47.35	0.000	**
	Neem, Nishyinda and turmeric	1600±58.56		
Weight gain (g)	Control	3200±35.49	0.000	**
	Neem, Nishyinda and turmeric	3100±52.29		
FCR	Control	2.20		
	Neem, Nishyinda and turmeric	1.93		

Table 2. Dressing percentages, relative giblet weight (heart, gizzard, liver and spleen) and pancreas weight of broilers supplemented with or without neem leaf, nishyinda leaf and turmeric rhizome from 1-5 weeks of age

Variables	Treatments	Average weight (Mean ± SEM)	P value	Significance level
Dressing percentage	Control	64.400±0.404	0.939	**
	Neem, Nishyina and Turmeric	64.470±0.961		
Relative heart weight (gm)/chicks	Control	0.420±0.032	0.002	**
	Neem, Nishyina and Turmeric	0.511±0.032		
Relative gizzard weight (gm)/chicks	Control	1.460±0.034	0.606	**
	Neem, Nishyina and Turmeric	1.440±0.014		
Relative liver weight (gm/chicks	Control	2.530±0.034	0.001	**
	Neem, Nishyina and Turmeric	2.610±0.032		
Relative spleen weight (gm)/chicks	Control	0.120±0.011	0.011	**
	Neem, Nishyina and Turmeric	0.130±0.015		
Relative pancreas weight (gm)/chicks	Control	0.230±0.011	0.001	**
	Neem, Nishyina and Turmeric	0.25±0.017		

**= Significant at 1% level of probability (0.00-0.01); NS = Not significant (≥ 0.05)
Relative weight (g) = Weight of organ/Live body weight of bird × 100; Dressing % = Dress weight of bird/Live weight of bird

Table 3. Data Showing Economics of Broiler Production Kept Under Treatment and Control Groups from Day Old to 5 Weeks of Age

Description	Group-A	Group-B
Cost/chick(Taka)	30.00	30.00
Average feed consumed (Kg)/Chicks	3.200	3.100
Feed price/kg (Taka)	40.00	40.000
Cost of herbal growth promoters (Taka)	0.00	2.00
Feed cost (Taka)	128.00	124.00
Miscellaneous (Taka)	20.00	20.00
Total cost/broiler (Taka)	178.00	176.00
Average live weight (Kg)	1.450	1.600
Sale price/Kg live wt. (Taka)	140.00	140.00
Sale price/broiler (Taka)	203.00	224.00
Net profit/broiler (Taka)	25.00	48.00
Profit/Kg live weight (Taka)	17.24	30.00

Study of Neem Leaf, Nishyinda Leaf and Turmeric Rhizome on Hematological Parameters of Broiler

Observation of hematological parameter (RBC, Hb, PCV, ESR) on 7th day and 35th day did not show any significant difference ($P < 0.05$) between the control and Neem, Nishyinda leaves and Turmeric treated groups.

Table 4. Study of Neem Leaf, Nishyinda Leaf and Turmeric Rhizome on Hematological Parameters of Broiler

Days	Blood parameters	Treatments	Average parameters (Mean±SEM)	blood value	P value	Significance level
7th Days	TEC	Control	192.39±1.037		0.000	**
		Neem, Nishyinda and turmeric leaf	199.29±0.992			
	Hb gm %	Control	6.00±0.089		0.337	**
		Neem, Nishyinda and turmeric leaf	6.37±0.438			
	PCV %	Control	17.35±0.599		0.011	**
		Neem, Nishyinda and turmeric leaf	17.97±0.456			
	ESR Mm/in 1st hour	Control	10.68±0.316		0.000	**
		Neem, Nishyinda and turmeric leaf	8.78±0.316			
35th Days	TEC Million/m m3	Control	247.67±1.028		0.000	**
		Neem, Nishyinda and turmeric leaf	298.39±0.751			
	Hb gm%	Control	6.92±0.491		0.241	**
		Neem, Nishyinda and turmeric leaf	7.79±0.111			
	PCV %	Control	18.00±0.134		0.000	**
		Neem, Nishyinda and turmeric leaf	19.95±0.022			
	ESR mm/in 1st hour	Control	7.40±0.268		0.004	**
		Neem, Nishyinda and turmeric leaf	5.24±0.554			

DISCUSSION

Addition of herbal growth promote Neem, Nishyinda leaves and Turmeric Rhizome improved the weight gain of the broilers in this study. These results are in line with the findings of Misra and Singh (2000), who reported that higher weight gain in broilers, drinking water supplemented with Nishyinda and Turmeric extract. on Treatment 5 utilized their diets better which was evidenced by their higher carcass values. The percentages of back, wing, head and shank did not differ significantly across the treatments more so, these parts carry less value in terms of meat yield and consumer preference. The values for liver, heart, gizzard and lungs did not differ significantly, this could be linked to the absence of anti-nutritional factors in the diets, because higher physiological activities by these organs is triggered by the presence of anti-nutritional factors and their concomitant effect. In conclusion, results of the present study showed that supplementation of diet with 1% (NLM+PLM) improve performance, feed utilization, dressing percentage and carcass yield therefore this combination of neem leaf meal and pawpaw leaf meal can serve as an effective replacement for chemical based growth promoters in broiler production.

The birds supplied drinking water supplemented with herbal growth promoters, Neem, Nishyinda leaves and Turmeric rhizome utilized their feed more efficiently than those supplied drinking water without addition of the growth promoters. These results are in line with the findings of Ahmad (2005), who reported higher weight gain in broilers fed rations supplemented with garlic. The use of Neem, Nishyinda leaves and Turmeric rhizome extract with drinking water showed more increase in live weight of the birds as compared to control group in this study, which is also in agreement with the findings of Samanta and Dey (1991), who concluded that powdered garlic may be incorporated as a growth promoter in the ration of Japanese quails.

Supplementation of Neem, Nishyinda and Turmeric extract did not exhibit any effect on the dressing percentage values of the broilers in this study. The results of the present study are in line with those observed by Ahmad (2005), who reported a non-significant effect on broiler dressing percentage values due to the inclusion of garlic in the diet of broilers.

Inclusion of 1% Neem, Nishyinda and Turmeric extract with drinking water exerted significant effect on the mean relative heart, gizzard, liver, spleen, pancreas weights of the broilers used in this study. Use of various levels of herbal growth promoters in the rations exhibited an increase in the profit margin of the broilers as compared to those using ration without the addition of these growth promoters. Supplementation of 1% Neem, Nishyinda and Turmeric with drinking water was found to be more profitable than without garlic supplementation in broiler rearing. The results of the present study are in line with the findings of Ahmad (2005), who reported that dietary inclusion of garlic in the rations was more beneficial in broiler production.

Increase in the profit margin of the birds supplied drinking water containing herbal growth promoters may be attributed to the better efficiency of feed utilization, which resulted in more growth and better feed to gain ratio, ultimately leading to higher profit margin in the broilers reared on Neem, Nishyinda and Turmeric supplemented drinking water. Growth promoting effects in broilers by using medicinal plants was reported earlier in Bangladesh by Nath et al. (2012).

Supplementation of Neem & Nishyinda leaves and Turmeric in the treatment group caused improvement in the feed efficiency as compared to that of control group. Birds supplemented with Neem, Nishyinda leaves and Turmeric had higher body weight, weekly gain weight, feed consumption and feed efficiency.

CONCLUSION

It is concluded that supplementation of Neem, Nishyinda leaves and Turmeric Rhizome of treatment groups caused significant increase in live body weighed and improvement in weekly gain in weight and feed efficiency as compared to that of control group of broilers. The results of the present study may be due to antimicrobial and anti-protozoal properties of Neem, Nishyinda and Turmeric which help to reduce the microbial load of birds and improved the feed consumption and feed efficiency of the birds. Thus, polyherbal supplementation in broiler rations may be useful for the production of broiler as an alternative to commercial growth promoters.

REFERENCES

1. Ahmad S, 2005. Comparative efficiency of Turmeric, garlic, Cinnamon and kalongi as growth promoter in broiler. M.Sc. (Hons.) Thesis, Department Poultry Sciences, University of Agriculture, Faisalabad, Pakistan. 2. Banerjee GC, 1998. A Text Book of Animal Husbandry. 2nd edition. India publication, Delhi.

2. Chang HW, 1995. Antibacterial effect of spices and vegetables. Food Industries, 27: 53-61.

3. Chowdhury NY, Islam W and Khalequzzaman M, 2009. Insecticidal activities of the leaves of nishyinda (vitex negundo verbinaceae) against tribolium castaneum hbst. Pakistan Entomologist, 31: 25-31.

4. DD Nath, M. M. Rahman, F Akter and M Mostofa, 2012. Effects of tulsi, black pepper and cloves extract as a growth promoter in broiler. Bangladesh Journal of Veterinary Medicine, 10: 33-39.

5. DeMan JC, 1975. The probability of most probable numbers. European Journal of Applied Microbiologyand Biotechnology, 1: 67-78.

6. Endtz HP, Rujis GH and Van Klingeren B, 1991. Quinolon resistance in Campylobacter isolated from man and poultry following the introduction of fluroquinolones in veterinary medicine. Journal of Antimicrobial Chemotherapy, 27: 199-208. 8.

7. Ghosh TK, 1992. Efficiency of liver herbal product on counteracting flatoxin on broiler birds. Indian Poultry Review, 32: 33–34.

8. Hileman B, 2002. Debate over health hazards of putting antibiotics in animal feed heats up in the USA, Chemical and Engineering news. 1999.
   Available at: http://www. organiconsumers.org/Toxic/bioticsinfeed.cfm Accessed August 2.

9. Juven BJ, Kanner J, Schved F and Weisslowicz H, 1994. Factors that interact with the antibacterial action of thyme essential oil and its active constituents. Journal of Applied Bacteriology, 76: 626-631.

10. Kale BP, Kothekar MA, Tayade HP, Jaju JB and Mateenuddin M, 2003. Effect of aqueous extract of Azadirachta indica leaves on hepatotoxicity induced by antitubercular drugs in rats. Indian Journal of Pharmacology 35: 177-180 16.

11. Kumar OM, 1991. Effect of Liv-52 syrup on broiler performance in North Eastern Region. Indian Poultry Review, 22: 37–38.

12. Lin CC, Wu SJ, Chang CH and Nu LT, 2003. Antioxidant activity of Cinnamomum cassia. Phytotherapy Research 17: 726-730.

13. Manwar SJ, Thirumurgan P, Konwar D, Chidanandaiah and Karna DK, 2005. Effect of Azadirachta indica leaf powder supplementation on broiler performance. Indian Veterinary Journal, 84: 159-162. 20.

14. Mastura M, Nor Azah MA, Khozirah S, Mawardi R and Manaf AA, 1999. Anticandidial and antidermatophic activity of Cinnamomum species essential oils. Cytobios, 98: 17-23.

15. MR Molla, MM Rahman, F Akter and M Mostofa, 2012. Effects of Nishyinda, black pepper and cinnamon extract as growth promoter in broilers. The Bangladesh Veterinarian, 29: 69 – 77.

16. Samanta AR and Dey A, 1991. Effect of feeding garlic (*A. sativum Linn.*) as a growth promoter in Japanese quails (*C. coturnix japonica*) and its influence on dressing parameter. Indian Journal of Poultry Science, 26: 42–145.

17. Tipu MA, Akhtar MS, Anjum MI and Raja ML, 2006. New dimension of medicinal plants as animal feed, Pakistan Veterinary Journal, 26: 144-148.

13

A STUDY ON PREVALENCE OF PESTE DES PETITS RUMINANT (PPR) IN GOAT AT BAGMARA UPAZILLA AT RAJSHAHI DISTRICT IN BANGLADESH

Md. Mostafijur Rahman[1*], Khondoker Jahengir Alam[1], Md. Shah Alam[1], Md. Mahmudul Hasan[2] and Monalisha Moonmoon[2]

[1]Department of Pathology and Parasitology, Faculty of Animal Science and Veterinary Medicine, Patuakhali Science and Technology University, Babugonj, Barisal 8210; [2]Faculty of Animal Science and Veterinary Medicine, Patuakhali Science and Technology University, Babugonj, Barisal 8210, Bangladesh

Corresponding author: Md. Mostafijur Rahman; E-mail: mrrana@pstu.ac.bd

ARTICLE INFO

Key words

Prevalence,
PPR,
Black bengal goat

ABSTRACT

A study was conducted at Bagmara upazilla under the district of Rajshahi in Bangladesh during the period of 2 month from 20 September 2015 to 14 November 2015 to determine the prevalence of PPR in goat. A total of 72 diseases cases were recorded randomly irrespective of age, sex and breed over the study period. In the present study the overall prevalence of PPR in goat was 27.78%. In black Bengal goat, the prevalence of PPR in different age group i.e. up to 6 month, 7-12 month, 13-19 month, above 19 month were 18.18%, 42.30%, 22.22% and 12.50%, respectively. Over all, female black bengal goat (46.42%) was more infected than male (12.00%). The prevalence of PPR in different age group i.e. 7-12 month, 13-19 month were 42.85% and 25.00%, respectively; and males (16.66%) and female (23.07%) of Jamunapari goat were recorder. In general, black bengal goats have the higher prevalence (30.18%) than Jamunapari goat (21.05%). The non-immunized goat showed higher prevalence (38.77%) than immunized (4.34%) goat. All data were analyzed by using $\chi 2$ tests with P value at (P<0.05) level of significance.

INTRODUCTION

Small ruminant especially goat is very important in rural economy of the Bangladesh. Peste des petits ruminants (PPR) is a disease of major economic importance and imposes a significant constrain upon sheep and goat production owing to its high mortality rate (Asimet al., 2008). Peste des petits ruminant (PPR), which literally means "Plague of small ruminants", is an economically important disease of sheep and goats. PPR has been recognized as a highly contagious viral disease of small ruminants, particularly in goats in Bangladesh (Islam et al., 2001). The PPR virus is considered as one of the predisposing factor for respiratory disease complex in goats (Taylor et al., 1990). It is a highly contagious, infectious and fatal viral disease of domestic and small ruminants (Abdallalet al., 2012). In Bangladesh, the PPR virus was first identified during a severe outbreak in 1993 (Sil et al., 1995) which was further confirmed by world reference laboratory and found that the virus has a close relation with Indian isolates (West Bengal) of PPR virus at a cluster with Asian group (Barrett et al., 1997). The Peste des petits ruminant's virus (PPRV) has been classified under the family Paramyxoviridae, Order Mononegavirales and Genus Morbillivirus (Toberet al., 1998). As members of the family *Paramyxoviridae*, the PPR virus has an envelope. The genome of PPR virus has single stranded RNA, approximately 16kb long with negative polarity (Haas et al., 1995). PPR in goat has been recorded in 1993 from the border belt areas of south western districts (Sathkhira, Jessore and Barguna) of Bangladesh and then spreads throughout the country. By the year 1995, it is assumed 75 percent of the district in Bangladesh is affected with PPR (Debnath, 1995). It has been reported that the Black Bengal goats were more susceptible (67.24%) to PPR than Jamunapari breed (32.76%). Morbidity varies from 40-95% and mortality as high as 80-85% (Samad, 2000). Considering the above circumstances and present situation the study was conducted to study the prevalence (%) of PPR of goat at Bagmara upazila of Rajshahi district in Bangladesh.

MATERIALS AND METHOD

Study area and duration
The present study was conducted at Bagmara upazilla Veterinary hospital under the district of Rajshahi in Bangladesh. The study was conducted from 20th September, 2015 to 14th November, 2015.

Sample and sample size
The study was conducted on naturally PPR suspected goat brought to the veterinary hospital during the study period. A total number of 72 diseases cases were recorded during the study period. Samples were selected randomly irrespective of age, sex and breed over the study period. During the study period different aged goats are consider. Both Black Bengal and Jamunapari goat were grouped into 4 different age groups. Such as, up to 6 month, 7-12 month, 13-19 month, above 19 month of age goat were categorized as group I, II, III and group IV, respectively.

Data collection
The data were directly collected from the owner. Data were based on , Client/owners complains, anamnesis of patient (goat), clinical history, physical examination data (Inspection, Temperature, auscultation, respiration) and clinical sign of suspected goats. History of the cases were taken carefully from the owner based the following aspect (Age, Sex, Breed, Vaccination-yes/no, duration of rearing, housing, previous disease history etc).

Recording of signs and symptoms
Close inspection: Different exposed signs and symptoms were recorded carefully by close infection-Erosion of oral mucosa, respiratory distress, discharges from eyes, nose, mouth, rough coat, soiled hind quarter.

Temperature: Temperatures were recorded by indirect palpation per rectum by thermometer of every case and tabulate.

Indirect auscultation: Indirect auscultation was performed to hear the lung and tracheal sound to coincide with the symptoms of pneumonia.

Skin fold test: Skin fold test were performed to take the rough estimation of the degree of dehydration.

Diagnosis: Among all diseased cases of goat brought to the veterinary hospitals for treatment, presumptive diagnosis of PPR in goat were made on the basis of owners complains, clinical history, clinical signs.

Data analysis

The raw data were collected from owners complain, clinical history and clinical signs and recorded into a previously formed data record sheet. Positivity and prevalence rate were calculated by using MS excel programme. $\chi 2$ test and P value were calculated by the help of SSS (online) software. A descriptive analysis was performed to interpret the data.

RESULT AND DISCUSSION

The present study was conducted to observe the prevalence of PPR in goat. The prevalence and associated results of this study are summarized in the table given below.

Table 1. Over all prevalence of PPR in goat

Total Case	No. of infected goat	Percent (%)
72	20	27.78

In the present study the overall prevalence (%) of PPR was 27.78% (Table 1). Khan *et al.,* (2007) reported the overall prevalence of PPRV was 43.33% of the ruminant population in Punjab. They also mentioned the overall PPR antibody seroprevalence in goats was 39.02% which is significantly higher. Abubakar et al., (2008) recorded the prevalence of PPR in small ruminants in Pakistan was 40.98%, but overall was 46.7%. On the contrary in Cameroon, (N = 320), 35% PPR antibodies while for Nigeria (N= 382), the values was 56.5 (Majiyagbe et al., 1992).

Prevalence of PPR on the basis of age

The prevalence of PPR in goats (Black Bengal and Jamunapari) were high in the age between 7-12 months, the prevalence per cent were observed 42.30%, 42.85% in Black Bengal and Jamunapari, respectively. On the other hand, in adult goat (aged over 19 months) prevalence of PPR were quiet negligible; it was 12.1% in Black Bengal whereas no clinical case was recorded in case of Jamunapari goat during the study period (Table 2). The result was statistically not significant. The findings of this study were agreed with Venkataramanan J. (2005) and Blood et al., (1995) they found that more prevalence of PPR in goats were under 1 year of age, specially 4-12 months of age. Taylor et al., (1990) also reported the susceptibility of young animals aged 3 to 18 months was proved to be very high. Radostits et al., (2000) and Singh et al. (2004) also assessed that most prevalent in the goats less than one year.

The present study revealed that, females were more susceptible to PPR than male. The prevalence (%) was found 46.42% and 23.07% in female followed by 12.00% and 16.66% in male Black Bengal and Jamunapari goat, respectively (Table 3). The statistical data analysis shows that, the results of Black Bengal goat was significant where as the result obtained in case of jamunapari goat was statistically not significant (P<0.05). The actual cause is not known but it was assumed that, females were normally immunologically weaker than male due to some hormonal effects, Pregnancy or milking status (Chakrabarti, 2004). Samad, (2001) reported that around 60.23% female goat affected with PPR. But this study shows lower value than Samad's observation. This is might be due to Sample size and duration of the present study was lower than that of Samad (2000).

Table 2. Age wise prevalence of PPR in Black Bengal (BB) and Jamunapari goat

Black Bengal Goat					
Age group	No. of goat	No. of infected	Prevalence (%)	χ2 test	P Value <0.05
Group I	11	2	18.18		
Group II	26	11	42.30	2.2434	0.523455
Group III	9	2	22.22		
Group IV	8	1	12.50		
Jamunapari Goat					
Age group	No. of goat	No. of infected in PPR	Prevalence (%)	χ2test	P Value <0.05
Group I	3	0	-		
Group II	7	3	42.85	2.8145	0.421123
Group III	4	1	25.00		
Group IV	5	0	-		

Sex wise prevalence of PPR

Table 3. Prevalence of PPR in different sexes of Black Bengal goat

Black Bengal Goat					
sex	No. of goat	No. of affected in PPR	Prevalence (%)	χ2test	P Value <0.05
Female	28	13	46.42	4.1167	0.042463
Male	25	3	12.00		
Jamnapari goat					
Sex	No. of goat	No. of affected in PPR	Percent in affected	χ2test	P Value <0.05
Female	13	3	23.07%	0.0676	0.794935
Male	6	1	16.66%		

Breed wise prevalence of PPR

In the study area during the study period Black Bengal (BB) goat and Jamunapari goat were brought to the hospital for treatment. So this two breed were consider for the present study. The result of the breed wise prevalence of PPR was summarized and presented in (Table 4). The result of the present study revealed that, Black Bengal breed of goat have the higher prevalence rate (30.18%) than Jamunapari (21.05%) with χ2 test (0.3407) and P value (0.559402) at (P<0.05) level of significance.

Table 4. Breed wise prevalence of PPR in goat

Breed	No. of case recorded	No. of PPR case	Prevalence (%)	χ^2 test	P Value <0.05
BB	53	16	30.18		
Jamunapari	19	4	21.05	0.3407	0.559402
Total	72	20	27.78		

The statistical analysis of the results shows that the result was not significant (P<0.05). This result is in agreement with the Shailaet $al.,$ (1989) and Samad (2001) reported Black Bengal goats are more susceptible (67.24%) to PPR than Jamunapari breed (32.76%). Higher incidence of PPR in to indigenous Black Bengal goats (27.13%) may be due to enhance participation in disease surveillance and immune-suppression and irregular vaccination program (Mondalet $al.,$ 1995).

Prevalence of PPR on the basis of immune status

Table 5. Prevalence of PPR on the basis of immunological status irrespective of breed

Immune status	No. of goat	No. of affected in PPR	Percent infected	in χ^2 test	P Value <0.05
Vaccinated	23	1	4.34%		
Non-vaccinated	49	19	38.77%	5.8934	0.15197

The findings of present study showed that, the prevalence of PPR was higher in non-vaccinated 38.77 (%) as compared to vaccinated 4.34 (%) goat irrespective to breeds (Table 5). The statistical analysis revealed that the result was not significant with χ^2 test (5.8934) and P value (0.15197) at (P<0.05) level of significance. This result supports the earlier report where higher prevalence of PPR (68.38%) was found in the unvaccinated goat (Gibbs et al., 1979).

REFERENCES

1. Abdallal AS, Majok, and Malik K, 2012. Sero-prevalence of peste des petits ruminant's virus (PPRV) in small ruminants in Blue Nile, Gadaref and North Kordofan states of Sudan. Journal of Public Health and Epidemiology, 4: 59-64.
2. Abubakar M, Jamal SM and Hussain M Ali, 2008. Incidence of peste des petits ruminants (PPR) virus in sheep and goat as detected by immuno-capture ELISA (Ic ELISA). Small Ruminant Research, 75: 256-259.
3. Asim M, Rashid A, and Chaudhary AH, 2008. Effect of various stabilizers on titre of lyophilized live attenuated Pest des petits ruminants (PPR) vaccine. Pakistan Veterinary Journal, 28: 203-204.
4. Barrete T, Pronab D, Sreenivasa BP and Corteyn M, 1997. Recent epidemiology of peste des petitsruminants virus (PPRV). Veterinary Microbiology, 88: 125-130.
5. Blood DC, Rodostis OM and Gay CC, 1995.Veterinary Medicine. 8th ed. BallierTindall, UK, pp. 837.
6. Chakrabarti A, 2004. A Text Book of Preventive Veterinary Medicine. 2ndedn. Kalyani Publisher, New Delhi, India, pp: 72-76.
7. Debnath NC, 1995. Peste des petits ruminants (PPR); an overview proceeding of the BSVER Symposium on Eradication of Rinderpest and Related Diseases. 2nd December, Dhaka, pp: 9-13.
8. Gibbs EPJ, Taylor WP and Lawman MJP, 1979. Classification of Peste des petits ruminants (PPR) virus as the fourth member of the genus Morbilivirus. International Journal of virology, 15:35-41.

9. Haas TJ, 1995. PPR virus as the fourth member of the genus Morbilivirus. International Journal of Virology, 16: 32-53.

10. Islam MR, Shamsuddin M, Das PM and Dewan ML, 2001. An outbreak of Peste des petits ruminants in Black Bengal goats in Mymensingh, Bangladesh. The Bangladesh Veterinarian, 18: 14-19.

11. Khan HA, Siddique M, Arshad MJ, Khan QM, and Rehman SU 2007. Sero-prevalence of peste des petits ruminants (PPR) virus in sheep and goats in Punjab province of Pakistan. Pakistan Veterinary Journal, 3: 109-112.

12. Majiyagbe KA, Shamaki D, Kulu D, and Udeani TKC, 1992. Peste des petite ruminants (PPR) and rinderpest (RP) antibodies in clinically normal small ruminants in the Cameroon and Nigeria. Small ruminant research and development in Africa, Proceedings of the Second Biennial Conference of the African   Small Ruminant Research Network AICC, Arusha, Tanzania 7-11 December 1992.

13. Mondal AK, Chottopadhuay AP, Sarkar SD, Saha GR and Bhowmik MK, 1995. Report on epizootological and clinico-pathological observation on peste des petits ruminants (PPR) in West Bengal. Indian Journal of Animal Health Bulletin, 64: 261.

14. Radostis OM, Blood DC, and Gay CC, 2000. Veterinary Medicine, 9thedn., Bailliere and Tindall, London, pp:1059-1063.

15. Samad MA, 2000. PoshuPalon O Chikitsavidya (Animal Husbandry and Medicine). 2ndEdn., Published by M. Bulbul, Bangladesh Agricultural University campus, Mymensingh, Bangladesh.

16. Samad MA, 2001. PoshuPalon O Chikitsavidya (Animal Husbandry and Medicine). 2ndEdn., Published by M. Bulbul, BAU campus, Mymensingh, Bangladesh.

17. Shaila MS, Purushothaman V, Bhavasar D, Venugopal K, and Venkatesan RA, 1989. Peste des petits ruminants of sheep in India. Veterinary Record, 125:602.

18. Sil BK, Rahman MF, and Taimur MJF, 1995. Observation on outbreaks of peste des petits ruminants in organized goat farms and approach of its treatment and preversation Second BSVER Annual Scientific Conference. Programme and Abstracts. Dhaka, 2- 3 December, pp.10.

19. Singh RP, Sreenivasa BP, Dhar P and Bandyopadhyay SK, 2004. A sandwich-ELISA for the diagnosis of Peste des petits ruminants (PPR) infection in small ruminants using anti-nucleocapsid protein monoclonal antibody. Archive of Virology, 149: 2155-2170.

20. Taylor WP, 1990. Epidemiology of peste des petits ruminants. Preventive Veterinary Medicine, 3: 135-197.

21. Tober C, Seufert M, Schneider H, Billeter MA, Johnston IC, Niewiesk S, terMeulen V, Schneider-Schaulies S, 1998. Expression of measles virus V protein is associated with pathogenicity and control of viral RNA synthesis. Journal of Virology, 72: 8124–8132.

22. Venkataramanan J, 2005. The isolation of peste des petits ruminant's virus from Northern India. Veterinary Microbiology, 51:207-212.

14

ON-FARM WELFARE ASSESSMENT OF DAIRY CATTLE BY ANIMAL-LINKED PARAMETERS IN BANGLADESH

Shamir Ahsan, M. Ariful Islam* and Md. Taohidul Islam

Department of Medicine, Faculty of Veterinary Science, Bangladesh Agricultural University, Mymensingh-2202, Bangladesh

*Corresponding author: M. Ariful Islam; E-mail: maislam77@bau.edu.bd

ARTICLE INFO	ABSTRACT
Key words Animal-based parameters, Welfare assessment, Dairy cattle, Bangladesh	The present study was conducted in 33 dairy farms to evaluate the welfare quality of Australian-zebu cross bred cows through some animal-based welfare indicators. The main aims of this research were to identify welfare issues facing dairy cows and investigate whether indicators are associated with measures of welfare and performance efficiency. The assessment of animal welfare was performed (330 animals) Australian-cross breed in family dairy farms at Sirajganj district of Bangladesh. Data were collected through face-to-face interview with farmers, followed by an inspection and observation of dairy cows. A total 330 females (43 heifers and 287 cows) were included in this study. Body condition, body cleanliness, injury, lameness, health status and milk yield were assessed. Among studied animals, body condition score 2 about (65.5%), hock joint injury (83.6%), knee injury (48.8%), and a pronounced state of poor cleanliness on: dirty udder (55.9%), flank (55.0%) and hind limbs (96.4%) were observed. Health status including diarrhea, respiratory distress, coughing, nasal and ocular discharge were present in some animals. The results indicate that very good BCS and mastitis free cows are related to higher milk yields. Results of this study may indicate the some indicators that influence the animal welfare and productivity in selected farms. As this work was a preliminary study, so the comprehensive research is needed to further develop the prototype protocol.

INTRODUCTION

Animal Welfare (AW) has been defined by the World Organization for Animal Health (OIE) as the broad term used to describe how an individual is coping with the conditions in which it lives. The welfare of dairy cows encompass nowadays a major concern of public interest extending in most of the countries, due to its impact on health and productions of animals and, implicit, upon public health. Good animal health and welfare is an explicit goal of livestock farming. AW is a relatively new topic which is just beginning to attract attention in Asia including Bangladesh. The welfare of dairy cattle and the risk factors causing poor welfare of dairy cattle in Bangladesh have not previously been determined, hence the relevance and need for the current study. On-farm assessment of animal welfare can be based on the evaluation of the provision of resources and management, direct observation of the animals and examination of farm records (Whay et al., 2003). The basic principles of animal welfare are defined by both the physical health as mental, and include aspects such as absence of prolonged hunger and thirst, thermal comfort, the absence of injuries, inappropriate management-induced pain, diseases, social behavior and expression, human-animal relationship, etc. Thus, the Welfare Quality protocol based its assessment of animal welfare predominantly in animal based measures (e.g., behavior, and health). When this measure is not sensitive or applicable to check a criterion, measures based on the resources (e.g., installations) or in the management (for example, management procedures) are used.

Animal welfare is recognized as an essential component of the social pillar of sustainability for the dairy industry. The animals need to have their feed provision consistent with their needs, easy access to drinking bowls and troughs, and total freedom of movement (Butler and Smith, 1989). In addition, the environment should provide thermal comfort conditions for animal's sufficient size in the rest area, maintain standards of hygiene and cleanliness in order to avoid the proliferation of pathogenic microorganisms (Fonseca and Santos, 2000; Barkema et al., 1998). The increasing attention toward animal welfare has resulted in the formation of many different welfare assessment protocols, such as the newly developed European Welfare quality protocols and Animal Needs Index, both of which include many different welfare indicators. For dairy cattle cleanliness, skin lesions and injury in different body parts, lameness and milk yield recur as important factors contribution to the welfare of the cow or indicating its general condition (Brtussek et al., 2000; Welfare Quality, 2009). The farm animal welfare is provided especially by housing and breeding systems suitable for animal health and behavioral needs and by proper farming practices, as well (Broom, 2004). Welfare is a condition of the animal, not something transmissible to it, and it range between very poor and very good (Loberg and Lidfors, 2001; Broom, 2004). Assessment of animal welfare can be done by several methods. Thus, evaluation can be based on behavioral, physiological, psychopathological parameters or productive performances. All the indicators have inconveniences and, in this way, are not reliable, used as sole assessment techniques. For this reason it was suggested that better results could be obtained in measuring animal welfare by using a system of indicators instead of individual parameters (Winckler et al., 2003; Rousing et al., 2007). Usually, determination of these indicators requires expensive investigations or high experimental effort, therefore they are inadequate in field research. However, the data in this study were collected according to the validated European Welfare Quality protocol (Welfare Quality, 2009) for dairy cattle but in some extend the protocol was modified based on local management system.

In Bangladesh, dairy cows in different areas are subjected to production systems that are not friendly to their welfare status of dairy animals, therefore, it is needed to assess the welfare status of zero-grazed dairy cows. For the current study, we measured some anima-based indicators for evaluation of dairy cow welfare in farm level. Globally, AW is considered an important tool for farming of animals, but this issue is a relatively new attractive topic in Bangladesh. To our knowledge, there is no study has been conducted so far to assess the dairy animal welfare in Bangladesh. The present study was conducted in small holder dairy farms to evaluate welfare quality of Australian-zebu cross bred cows through some animal-based welfare indicators validated by the European project Welfare Quality® (Welfare Quality, 2009). The main aims of this research were to identify welfare issues facing dairy cows and investigate measures of welfare and performance efficiency.

MATERIALS AND METHODS

The present study was conducted during the period from February 2016 to April 2016 in Sirajganj district of Bangladesh. During the visit farm, the farmer were informed of the purpose of the study and assured that their participation was voluntary and their identity was kept confidential. The on-farm assessment took an hour long interview and direct observation on ocular discharge; nasal discharge, hampered breathing, diarrhea, body condition score, cleanliness of udder, flank/upper legs and lower legs and lameness. Production and health records were collected from the farmers and record books for each animal.

Selection of farms

The family dairy farms were selected in Sirajganj district for complete the research, as it is the major milk producing district in Bangladesh. In the case of Sirajganj, farms were defined by the criteria (a) location, (b) size and (c) production system so as to cover the farm types that make important contributions to milk production in the region. Nearly half of the milk in Bangladesh is produced on the northern region, where Sirajganj district is located.

Farms and animals

This study involved the collection of data from dairy farm. The study focused on the Australian cross breed, as the majority of dairy cow in Sirajganj. About 33 dairy farm located in Sirajganj, a sample of 330 females that responded to a questionnaire. The herd size of the farms represents minimum 10 female cow (lactating cow, dry cow and heifers). The dairy cattle rearing method was similar (artificial insemination, calves being separate from mother at the age of 1 to 7 days and fed by man) in all family hold farms.

Questionnaire

Face-to-face interviews with the farmers were carried out using a questionnaire with multiple-choice and semi-closed questions to collect animal-linked parameters related to welfare. The first part of the interview covered data on farm characteristics such as number of dairy cattle in each category (total cows, lactating cows and heifers), and total milk production. The second part of the questionnaire referred to the welfare assessment parameters such as BCS, body injury, lameness, poor cleanliness, most important disease (mastitis, reproductive problems, ocular and nasal discharge, diarrhea, or other). At all times, farmers had the opportunity to clarify questions and added personal information and remarks.

Body condition scoring

An average score of 2 is the most desirable for the majority of the herd. A four point body condition score system was used, in which a score of 0 was very thin, and a score of 3 was very fat. These are extreme scores and should be avoided. Because only extreme BCS (too thin or too fat) is likely to have negative effects on animal welfare. Cows were classified as BCS 0 = very thin, BCS 1= thin, BCS 2 = ideal, and BCS 3 = obese. A condition score of 2 is thought to be acceptable for lactating Australian-local cross bred dairy cows.

Cleanliness score

Experimental animals were scored using a modification of the system described by Krebs et al. (2001). A scoring chart divided the animal body into five identifiable areas which were rated on a scale with anchor points at each end (0: clean and 1: dirty). The five regions were: hind limb, Udder, Flanks, dull hair coat, thick and shiny hair coat.

Lameness score

Lameness is painful to the animal, it is a serious welfare issue as cows suffer and is costly to the dairy farm business. Locomotion scoring is based on the observation of cows standing and walking (gait), with special emphasis on their back posture. The lameness score was recorded after the afternoon milking using a scheme proposed by Breuer et al., (2000). Data were collected when cows were present in the shed. A score of 0 to 3 was used, where 0 was assigned when the animal was not lame (normal gait), 1 was given when the cow was mild lame, 2 was indicates, moderate lameness in cow and 3 was recorded when the cow was suffer from severely lame.

Milk yield

Data on milk traits (production) of per day were obtained directly from farmers' question and sore were from record book of farm. Cows which were complete 3 to 6 month lactation were considered. Manual milking was carried twice daily (at early morning and afternoon).

Skin Lesions and injury of body regions

During data collection on farm, cows were inspected for different lesions on different parts of animal body. The lesion and injuries was observed by directly and data was collected by observation of two side of body with some modification: the area around the carpal and tarsal joints, any lesions on head, abdomen and tail and percentage of cows per herd that presented skin lesion in each area was calculated (%).

Data Analysis

Percentage (%) of values was calculated for different variable. To find out the significance difference in milk yield in relation to history of mastitis and BCS, student's t test and analysis of variance were done by using SPSS v. 20.

RESULTS

Milk Yield

The mean values of milk yield are reported in Table 1, where milk yield categorized by BCS and mastitis (last 6 months from date of data collection). The average milk production of 173 cows was about 9.34% when cows were free from mastitis condition. On the other hand, high body scoring (3) animals produce higher amount of milk (average 10.32 L) and low body scoring (1) produce lower amount of milk (average 8.56 L). In below table 6, the significant level of mastitis absent cows is higher than mastitis present cows. In BCS case, there is no variation between score 1 and score 2 and milk production is higher than two other.

Table 1. Mean value (± SE) of milk yield

Measure	No. of animals		Milk Yield (Litre) (Mean ± Slandered Error)
History of Mastitis (last 6 Months)	Present	37	7.65 ± 0.42
	Absent	173	9.34 ± 0.21**
BCS	Score 1	48	8.56 ± 0.36 b
	Score 2	134	8.95 ± 0.23 b
	Score 3	28	10.32 ± 0.68 a
Total	210		9.04 ± 0.19

**=Significant at $p < 0.01$; Values with different letter differ significantly ($p < 0.05$)

Body condition Scoring (BCS)

Body condition is a subjective assessment of the amount of fat or amount of stored energy, a cow carries. In this study, majority of cows in all farms shows scoring 2. The highest percentage about 65.5% and lowest percentage about 10.0% those animals were represents scoring 2 and 3, respectively (Table 2).

Lameness

In that study, majorities of cows were not showed any lameness sign during on-farm observation. The higher lameness percentage about 94.8% that represents animals were free from suffering lameness condition. The percentage of mild, moderate and severe were 1.5, 3.0 and 0.6%, respectively (Table 3).

Table 2. Percentage value of cows according to BCS

Measure		No. of Animals	Percentage (%)
BCS	Score 1	81	24.5
	Score 2	216	65.5
	Score 3	33	10.0

Table 3. Percentage value of cows according to lameness

Parameter		No. of Animals	Percentage (%)
Lameness	Sound	313	94.8
	Mild	5	1.5
	Moderate	10	3.0
	Severe	2	0.6

Absence of injuries

In hock joint condition, in the present study most of the cows (83.6%) show healthy condition of the hock joint. The higher percentage of healthy hock about 83.6% and rest of the animals had swollen and mildly affected the hock joint 8.2% (Table 9). On the other hand during observation of knee injury, the highest percentage of no injury about 48.8%, swollen without skin injury about 19.4 and swollen with skin damage about 31.2% (Table 4).

Table 4. Percentage value of cows according to Leg injuries

Measure		No. of Animals	Percentage (%)
Hock Joint Condition	Healthy Hock	276	83.6
	Mildly affected Hock	27	8.2
	Swollen hock	27	8.2
Knee Injuries	No Injury	161	48.8
	Swollen without skin damage	64	19.4
	Swollen with skin damage	103	31.2

Cleanliness

In the present study, majorities of animal was showed dirty condition in different parts of body. The dirty percentage value in hind limbs, udder and flanks about 96.4, 55.9 and 55.0%, respectively (Table 5).

Health condition (Table 6)

In study, percentage of mastitis condition in farm is very lower about 12.6%. The hair loss and non-hock injuries percentage about 5.2 and 7.0%. Several symptoms highlighted an average percentage of cows with hampered respiration of 7.9% and nasal discharge, ocular discharge about 12.7 and 3.6%. Mean no. of coughs express per cow per 15 minutes and average percentage was about 3.3%. In feces condition most cows showed firm feces condition about 61.2% and other case not firm or liquid and liquid feces about 35.2, 3.6% respectively. About 76.4% cows showed normal teat and 23.0, 0.6% showed dry, acute lesion.

Table 5. Percentage value of different parts of cow's body according to cleanliness

Measure		No. of Animals	Percentage (%)
Dirty hind limbs	Present	317	96.4
	Absent	12	3.6
Dirty udder	Present	184	55.9
	Absent	145	44.1
Dirty flanks	Present	181	55.0
	Absent	148	45.0

Table 6. Percentage value of cows according to absence of disease

Measure		No. of Animals	Percentage (%)
Mastitis	Present	41	12.6
	Absent	289	87.6
Hair Loss	Present	17	5.2
	Absent	313	94.8
Non-Hock Injuries	Present	23	7.0
	Absent	307	93.0
Coughing	Present	11	3.3
	Absent	319	96.7
Nasal Discharge	Present	42	12.7
	Absent	288	87.3
Hampered Respiration	Present	26	7.9
	Absent	304	92.1
Ocular discharge	Present	12	3.6
	Absent	318	96.4
Feces	Firm	202	61.2
	Not Firm or Liquid	116	35.2
	Liquid	12	3.6
Teat Scoring Lactating	Normal	252	76.4
	Dry	76	23.0
	Acute Lesion	2	0.6

DISCUSSION

In this study farm were selected in Sirajganj district where large number of animals present and every family rear animals and those area should be selected to know the people should maintain or not maintain animal welfare in their farm. The 33 farms were small sample size and do not necessarily represent the welfare condition of cows throughout the Bangladesh. However the study probably constitutes the largest independently observe assessment of the welfare of dairy cows to have carried out in the Bangladesh. In this study, body scoring 2 animal were available in all farms. A body condition score of 2 may be the most desirable in late lactation. The prevalence of sore 2 is about 65.5% that indicates that those animals are suitable for farmers according to their income level. A condition score of 2 is thought to be acceptable for lactating dairy cow. Studer (1998) explained that high producing cows whose body condition score declines by 0.5 to 1.0 during lactation often experience anoestrus. Body condition affects productivity, reproduction, health and longevity of dairy cows. Dechow et al., (2001) found that higher body condition scores were favorably related genetically to reproductive performance during lactation. While higher body scores during lactation were moderately negatively related to milk production, both genetically and phenotypically. Good health is considered a prerequisite for welfare. Bovine lameness represents a major health problem for the dairy industry. A significant percentage of dairy cattle (59%) have severe lameness, this can be a sign of poor overall welfare standards within the herd. In the study the percentage of sound lameness is about (94.5%) that indicates the farmers were know the effect of lameness in their production and properly maintain their farm management. Hristov et al., (2008) noticed that lameness is indisputably the major welfare problem for the dairy cow. Lameness is a major welfare problem for dairy animals including pain and discomfort of long duration. This disease may be caused by several different factors, such as unbalanced nutrition, flooring and related time spent standing, etc. (Galindo et al., 2000; Winckler and Willen, 2001).

In this study lameness is less appear due farmers had awareness about lameness disease and some farmers kept their animals in pasture land in several times of day and vaccination should be done vaccination regularly. The causes of lameness in cows in the regions need to be investigated, considering that all herds have access to pasture for several hours per day, which is expected to reduce the incidence of the disease (Hernandez-Mendo et al., 2007). In this study dirty hind limb, udder, flanks are most common in all farms. The cows were spent several times in herd. Several dirty particles (such as feces, muds, urine) were present in herd that the cause dirtiness. Cows were several time lie down on floor that cause injury of body and body part show dirtiness. The prevalence of dirtiness was associated with surface of lying area and feeding different types of roughage. The percentage level of dirtiness in study area about 96.4%. The positive association between the prevalence of dirty hindquarters and head lunge impediments seems difficult to explain because more restricted stalls have often been associated with cleaner cows (Fregonesi et al., 2009). Most of the farmers used concrete floor in their farm that cause injury of knee region and few people used carpet or matt over the concrete floor where injury level was lower than concrete floor. The percentage of knee injury about 31.2% due to use of concrete floor and some associated factors. Barker et al., (2007) hypothesized that keeping cows temporarily in straw yards can thin the sole horn, which may lead to sole ulcers when cows are kept on hard floors after calving.

CONCLUSION

This study, the first done at dairy farm level in Bangladesh, is a first step towards finding a tool for veterinarians and farmers to assess dairy cow welfare. It is noted from our study that most of the farmers were not aware about the welfare issue related to dairy production. This study concluded that the most important hazards in relation to animal welfare were injury and dirtiness in different body parts, BCS and lameness. The reductions in productivity have been considered as an indicator of poor welfare indicators like BCS and milk yield was related to mastitis problem. It seems that lameness was the major welfare problem within the studied parameters. As this work was a preliminary study, it is, however, evident, that comprehensive research is needed to further develop the prototype protocol for the different production and housing system over the country.

ACKNOWLEDGEMENT

The authors are grateful to Dr. Pankaj Kumar Sen, Veterinarian, Milk Vita, Baghabari Ghat in Sirajganj and to farmers for their help and contribution in this research.

REFERENCES

1. Barkema HW, Schukken YH, Lam TJGM, Beiboer ML, Benedictus G and Brand A, 1998. Management practices associated with low, medium and high somatic cell count in bulk milk. Journal of Dairy Science, 81: 1917-1927.
2. Barker ZE, Amoy JR, Wright JI, Blowery RW and Green LE, 2007. Management factors associated with impaired locomotion in dairy cows in England and Wales. Journal of Dairy Science, 90: 3270-3277.
3. Bartussek H, Leeb C, Held S, 2000. Animal Needs Index for Cattle – ANI 35L/2000 cattle. Federal Research Institute for Agriculture in Alpine Regions BAL Gumpstein, A 8952 Irdning, of the Federal Ministry of Agriculture and Forestry, Austria.
4. Breuer K, Hemsworth PH, Barnett JI, Mathews LR and Coleman GJ, 2000. Behavioural response to humans and the productivity of commercial dairy cows. Applied Animal Behaviour Science, 66: 273-288.
5. Broom DM, 2004. Welfare. In Bovine Medicine: Diseases and Husbandry of Cattle. Ed. A.H. Andrews, R.W. Blowey, H.Boyd and R.G. Eddy 955-967, Oxford: Blackwell.
6. Butler WR and Smith RD, 1989. Interrelation between energy balance and postpartum reproductive function in dairy cattle. Journal of Dairy Science, 72: 767-772.
7. Dechow CD, Rogers GW and Clay JS, 2001. Heritability and correlations among body condition scores, production trait and reproductive performance. Journal of Dairy Science, 84: 266-275.
8. Fonseca LFL and Santos MV, 2000. Qualidade do leite e controle da mastite. Sao Paulo, Brasil, Ed. Lemos, 189.
9. Fregonesi JA, Von Keyserlingk A, Tucker CB, Veira DM and Weary DM, 2009. Neck-rail position in the free stall affects standing behaviour and udder and stall cleanliness. Journal of Dairy Science, 92: 1979-1985.
10. Galindo F, Broom DM and Jackson PGG, 2000. A note on possible link between behaviour and the occurrence of lameness in dairy cows. Applied Animal Behaviour Science, 67: 335-341.
11. Hernandez-Mendo O, Von Keyserlingk, Veira DM and Weary DM, 2007. Effects of pasture versus free stall housing on lameness in dairy cows. Journal of Dairy Science, 90: 1209-1214.
12. Hristov S, Stankovic B, Zlatanovic Z, Joksimovic-Todorovic MV and Davidovic V, 2008. Rearing condition, health and welfare of dairy cows. Biotechnology in Animal Husbandry, 24: 25-35.
13. Krebs S, Danuser J and Regula G, 2001. Using a herd monitoring system in the assessment of welfare. Acta Agriculturae Scandinavica, Section A - Animal Science, 30: 78-81.
14. Loberg J, Lidfors L, 2001. Effect of stage of lactation and breed on dairy cows' acceptance of foster calves. Applied Animal Behaviour Science, 74: 97-108.
15. Rousing T, Jakobsen IA, Hindhede J, Klaas IC, Bonde M, Sørensen JT, 2007. Evaluation of a welfare indicator protocol for assessing animal welfare in AMS herds: researcher, production advisor and veterinary practitioner opinion. Animal Welfare, 16: 213-216.
16. Studer E, 1998. A veterinary perspective of on evaluation of nutrition and reproduction. Journal of Dairy Science, 81: 872-876.
17. Welfare Quality® assessment protocol for cattle, 2009. Welfare Quality® Consortium, Lelystad Netherland, ISBN/EAN 978-90-78240-04-4,180 p. 2009
18. Whay HR, 2007. The journey to animal welfare improvement. Animal Welfare, 16: 117-122.
19. Whay HR, Main D, Green L and Webster AJF, 2003. Assessment of the welfare of dairy cattle using animal-based measurements: direct observations and investigation of farm records. Veterinary Record, 153: 197-202.
20. Winckler C and Willen S, 2001. The reliability and repeatability of a lameness scoring system for use as an indicator of welfare in dairy cattle. Acta Agriculturae Scandinavica, Section A-Animal Science, 51: 103-107.
21. Winckler C, Capdeville J, Gebresenbet G, Hörning B, Roiha U, Tosi M, Waiblinger S, 2003. Selection of parameters for on-farm welfare-assessment protocols in cattle and buffalo. Animal Welfare Volume, 12: 619-624.

15

EFFECT OF GENETIC AND NON-GENETIC FACTORS ON MILK PRODUCTION PERFORMANCE OF HOLSTEIN-FRIESIAN × LOCAL CROSSBREDS AT THE VILLAGES OF NOAKHALI DISTRICT IN BANGLADESH

Farukul Islam[1*], Muhammad Omar Faruque[2], Fowjia Ferdous[3], Sifat Hossain Joya[1], Rafiqul Islam[3] and Md. Shamsul Hossain[1]

[1]Bangladesh Agricultural University, Mymensingh-2202, Bangladesh; [2]Bangladesh Livestock Research Institute, Saver, Dhaka, Bangladesh; [3]Department of Livestock Services, Dhaka, Bangladesh

*Corresponding author: Farukul Islam; E-mail: farukkrishibid@gmail.com

TICLE INFO	ABSTRACT
	A total of 210 Holstein Friesian (HF) × Local crossbred cattle were examined to collect data like, test day milk production (MT), peak milk production (PM), lactation period (LD), green grass used the day before test milk production (GG), cost involved to feed the cow with concentrate feed on the day before test milk production (CP), age, body weight of cows (BW) and ancestry of test cows to define the grade. The data were collected using a pre-structured questionnaire at the villages of Noakhali district in Bangladesh during October to November 2016.The effect of grades, age, body weight, concentrate feed and green grass on milk production were evaluated. To study the effects, Duncan's Multiple Range Test and Pearson's correlation coefficient were performed using the Statistical Package for the Social Sciences. Positive correlation of MT with CP (0.794) and GG (0.453) were estimated. Ages of cows did not affect MT, PM and LD significantly. In grade two, the highest, MT (18.75±2.62 liter/cow/day) and PM (20.75±2.62 liter/cow/day) were reported for body weight group 3 while, the longest LD (219.88±0.47) was reported under body weight group 2. However, under grade three in body weight group 3, MT, PM and LD were 15.57±0.78 liter/cow/day, 18.00±0.78 liter/cow/day and 218.79±0.80 days/cow, respectively. Present study might be suggested that body weight group 3 under grade three was better for MT and PM while, body weight group 2 was better for LD but to come up with final decisions in this regard, further study addressing more numbers of crossbred cows with defined exotic blood percentages under different grades, seasons of calving, parity, feed and water management, housing, healthcare and farmers socio-economic status would be advisable. Finally, it might be indicated that grades, bodyweight groups, CP and GG affected milk production at the villages of Noakhali district in Bangladesh.
Key words Crossbred cattle Milk production Bangladesh	

INTRODUCTION

Huge gap between production and requirements of milk, influencing the importer to import milk in various forms from abroad though, Bangladesh has the greatest potentiality to increase milk production with proper utilization of its huge man power. It is a matter of hope for the nation that over the decades dairy sector has made tremendous improvements in the field of establishment of some crossbred dairy cattle genotypes and selected some improved local verities for milk production purpose and private entrepreneurs are involving in this promising sector day by day. Livestock is a vital component of agriculture and this is contributing more than 6% of total foreign exchange earnings and about 3.10% to gross domestic products (GDP) in Bangladesh (BER, 2015). In Bangladesh, per capita availability of meat and milk are 102 gm/day and 120 ml/day, respectively but per capita meat and milk requirement are 120 gm/day and 250 ml/day, respectively (BER, 2013), while, here in Bangladesh total annual requirement is 13.32 million tons (MT) of milk but annual production of milk is 3.46 million tons only (BER, 2012). In Bangladesh, improved varieties like: Pabna Cattle, Red Chittagong, Munshiganj Cattle, North Bengal Grey Cattle and introduced exotic breeds like Holstein-Friesian, Jersy, Sahiwal, Hariana, Sindhi, Australian, Sahiwal-Friesian, etc are available. Out of total cattle population (23.4 million) in Bangladesh, about 3.53 million milking cows, 2.61 million dry cows (cows without milk), 2.13 million draught cattle, and 4.20 million improved cattle were reported (Banglapedia, 2014). Digestibility of organic matter and crude protein were improved and the milk yield also increased by increasing the concentrate ratio in the diet from 30 to 45 percent but there was no effect on milk composition (Tuan, 2000). Feed intake, nutrient digestibility and live weight gain of indigenous cattle were improved through supplementation of straw-based diets with vigna hay, in Bangladesh (Hossain *et al.,* 2015). Friesian x Local crossbred cows's milk production performance under Bangladesh condition considerably improved over the decades (Bhuiyan, 2011) and this crossbred animals contributes about 24% of the 6.9 million breedable cows and heifers (Huque *et al.,* 2011). Among the various mating systems crossbreeding of local non-descript cattle with exotic breeds of high genetic potential, to improve genetic merit of the dairy animals, is considered to be a rapid and effective method (Usman *et al.*, 2012). To add proven bull in dairy cattle industry, progeny tested bulls for dairy development in the country are in progress (Bhuiyan *et al.*, 2015).Many research works have been carried out using Holstein Friesian and other exotic hi-yielding breeding bull through crossbreeding with local cattle to improve milk production but published data of systematic research works with specific grade number of Friesian X local crossbred cattle, assessments of the optimum amount of concentrate feed and green grass for maximum milk production at the rural villages in Bangladesh for this crossbred dairy cows is scanty. So, the present study was designed and conducted to explore knowledge about the milk production performances of Friesian X local crossbred cattle under different grades and to know the present status and relationship of concentrate feed and green grass use with milk production parameters at rural village condition in Bangladesh.

MATERIALS AND METHODS

Data like test day milk production (MT), peak milk production in a whole lactation period (PM), lactation period (LD) in days, green grass used the day before test milk production in kg per cow (GG), cost involved to feed the cow with concentrate feed on the day before test milk production per cow in Bangladeshi taka (CP), age of enumerated cows, body weight of cows (BW) and ancestry(dam and sire identity) of test cows to define the grade of the enumerated cows were collected. Door to door visits were performed to collect the said data using a pre-structured questionnaire from a total of 210 Holstein-Friesian X Local crossbred dairy cows at the villages of Sonaimuri, Senbagh, Suborno char and Noakhali sadar upazilas under the district of Noakhali in Bangladesh during October to November 2016. Numbers of Crossbred dairy cows of grade one, two and three were 5, 158 and 44, respectively. Experimental cows were divided into three body weight groups like below:

Body weight group one (BWG1): Body weight per cow under this group was in a range of 200 kg to below 300 kg.
Body weight group two (BWG2): Body weight per cow under this group was in a range of 300 kg to below 400 kg.
Body weight group three (BWG3): Body weight per cow under this group was in a range of 400 kg to 570 kg.

Grades were defined as follows:

Grade one: Progeny produced after successful mating between Local female and upgraded Friesian male (progeny produced after successful mating between upgraded Holstein-Friesian male and upgraded local female cattle).

Grade two: Progeny produced after successful mating between female of grade one and upgraded Friesian male.

Grade three: Progeny produced after successful mating between female of grade two and upgraded Friesian male.

All cows were inseminated artificially and all male calves were kept for meat purpose to sale. The following formulae were used to calculate **CPC** and **GGC**:

$$CPC = \frac{\text{Total concentrate feed cost spent per cow for test day milk production in BDT}}{Total\ milk\ production\ on\ test\ day\ per\ cow\ in\ liter}$$

$$GGC = \frac{\text{Total green grass supplied per cow for test day milk production in kg}}{Total\ milk\ production\ on\ test\ day\ per\ cow\ in\ liter}$$

The observation numbers of different traits were unequal and the design of the study was unbalanced factorial in nature. The recorded data were stored on to the excel spread sheet and edited for further analyses. Then data were analyzed for having means through compare means menu, to obtain the relationship among the traits like MT, CP and GG, Pearson's correlation coefficient were used through correlate menu, and Duncan's Multiple Range Test (DMRT) were used for performing mean comparisons using the Statistical Package for the Social Sciences version 14.0 (SPSS, 2005).

RESULTS

Correlation of MT with CP and GG

The correlations between different traits are presented in Table 1. Strong positive correlation (0.794) between MT and CP was observed in present study. Similarly, positive correlation (0.453) between MT and GG was also documented.

Effect of grades on MT, CPC and GGC

Table 2 shows the effect of grades on MT, CPC and GGC. Test day milk production was significantly varied among the grades of crossbred cows. The highest amount of MT was recorded for grade three (14.75±0.70 liter/cow) and this was followed by grade two and one. The lowest amount of CPC was found in grade three (9.99±0.70 BDT/liter), while the highest was noted in grade one. However, the highest GGC was reported in grade one (0.82±2.00 kg/liter/cow) and the lowest were in grade three.

Effect of grades on PM and LD

The effects of different grades on PM and LD are summarized in Table 3. The PD and LD were affected by grades significantly. The highest peak milk productions were recorded for grade three (17.11±0.70 liter/cow/day) and the lowest were reported for grade one (6.60±2.00 liter /cow/day). On the contrary, the LD between grades two and three did not vary significantly though the highest LD was recorded for grade three (217.98±0.72 days).

Effect of age groups of cow on MT, PM and LD

Table 5 briefs the effects of ages of cows under study on MT, PM and LD. Though the higher MT were reported for age group 3 (cows include in this group were of above 6 years to 10 years of old) than age group 1 (cows include in this group were of 2.5 years to 4 years of old) and age group 2 (cows include in this group were of above 4 years to 6 years of old). Similarly, the higher PM was reported for age group 3 than age group 1 and age group 2. On the contrary, LD was higher in age group 2 than 1, though the same were higher in age group 3 than 2. However, the effect of age group on MT, PM and LD was not significant.

Table 1. Correlations between MT and CP; MT and GG

Traits	MT
CP	0.794** (210)
GG	0.453** (209)

Note: CP-Concentrate feed cost in Bangladeshi Taka (BDT) to feed the cow for the test day milk production, GG-green grass in kg supplied per cow to produce milk on test day, MT-test day milk production in liter/cow, ** correlation is significant at the 0.01 level (2 tailed)

Table 2. Effect of grades on test day milk production and feed cost of Holstein-Friesian X Local crossbred dairy cattle genotypes

Parameter	Grade 1	Grade 2	Grade 3	F
MT	6.00^c±2.00 (5)	11.75^b±0.38 (158)	14.75^a±0.70 (44)	18.289
CPC	12.00^c±2.00 (5)	10.13^b±0.38 (158)	9.99^a±0.70 (44)	1.827
GGC	0.82^c±2.00 (5)	0.38^a±0.38 (158)	0.47^b±0.70 (44)	15.187

Note: MT-test day milk production in liter, CPC- cost in Bangladeshi Taka (BDT) to feed the cow for concentrate feed for per liter of milk production on test day, GGC-green grass in kg supplied to the cow to produce one liter of milk on test day, F-F value in one way ANOVA during post-hoc analysis for DMRT and ^{abc}Means with the different superscripts differed significantly within the column (P<0.05)

Table 3. Effect of grades on peak milk production and lactation period of Holstein-Friesian X Local crossbred dairy cattle genotypes

Parameter	Grade 1	Grade 2	Grade 3	F
PM	6.60^c±2.00 (5)	13.70^b±0.38 (157)	17.11^a±0.70 (44)	22.953
LD	190.00^b±2.00 (5)	217.58^a±0.39 (153)	217.98^a±0.72 (42)	7.197

Note: PM-peak milk production per day in liter, LD-lactation period in one calving. F-F value in one way ANOVA during post-hoc analysis for DMRT and ^{abc}Means with the different superscripts differed significantly within the column (P<0.05)

Table 4. Effect of age of cow on milk production performances of Holstein-Friesian X Local crossbred dairy cattle genotypes

Parameter	(AG1) 2.5 to 4 years	(AG2) above 4 to 6 years	(AG3) above 6 to 10 years	F
MT	12.37±0.69 (46)	11.81±0.47 (102)	12.93±0.61 (60)	1.489
PM	14.28±0.70 (45)	13.77±0.47 (102)	15.14±0.61 (60)	1.944
LD	214.29±0.70 (45)	216.69±0.48 (99)	219.74±0.62 (57)	1.392

Note: MT-test day milk production in liter, PM-peak milk production per day in liter, LD-lactation period in one calving. F-F value in one way ANOVA during post-hoc analysis for DMRT, AG1-age group 1, AG2-age group 2 and AG3-age group 3

Effect of body weight groups (BWG) on milk production performances under different grades

Grade one

Table 5 states the milk production performances of crossbred cattle under grade 1. All cattle were   with a body weight range like 200 kg to below 300 kg/cow under grade one crossbred cattle. Test day milk production and peak milk production per cow were 6.00±2.00 liter/cow/day and 6.60±2.00 liter/cow/day, respectively. Mean lactation period in days were 190.00±2.00/cow.

Grade two

Effects of BWG on MT, PM and LD in grade two are presented in Table 6. In grade two, MT, PM and LD were affected by body weight groups significantly. The higher MT (18.75±2.62 liter/cow/day) and PM (20.75±2.62 liter/cow/day) were reported for BWG3 than BWG2 and BWG1. But very interestingly, LD was reported higher under BWG2 than BWG1 and BWG3.

Grade three

Effects of BWG on MT, PM and LD in grade three are presented in Table 7.In grade three, MT, PM and LD were not affected by body weight groups significantly. MT, PM and LD were reported for BWG3 under grade three, 15.57±0.78 liter/cow/day, 18.00±0.78 liter/cow/day and 218.79±0.80 days/cow, respectively.

Table 5. Effect of body weight of cow in grade one on milk production performances of Holstein-Friesian X Local crossbred dairy cattle genotypes

Parameter	BWG1	BWG2	BWG3	F
	(200 to below 300 kg/cow)	(300 to below 400kg/cow)	(400 to 570kg/cow)	
MT	6.00±2.00 (5)	-	-	-
PM	6.60±2.00 (5)	-	-	-
LD	190.00±2.00 (5)	-	-	-

Note: MT-test day milk production in liter, PM-peak milk production per day in liter, LD-lactation period in one calving

Table 6. Effect of body weight of cow in grade two on milk production performances of Holstein-Friesian X Local crossbred dairy cattle genotypes

Parameter	BWG1	BWG2	BWG3	F
	(200 to below 300 kg/cow)	(300 to below 400kg/cow)	(400 to 570kg/cow)	
MT	8.91^c±0.69 (46)	12.69^b±0.46 (108)	18.75^a±2.62 (4)	46.455
PM	10.71^c±0.69 (46)	14.73^b±0.46 (107)	20.75^a±2.62 (4)	41.276
LD	213.82^b±0.70 (45)	219.88^a±0.47 (104)	200^c (4)	4.159

Note: MT-test day milk production in liter, PM-peak milk production per day in liter, LD-lactation period in days in one calving. F-F value in one way ANOVA during post-hoc analysis for DMRT and [abc]Means with the different superscripts differed significantly within the column (P<0.05)

Table 7. Effect of body weight of cow in grade three on milk production performances of Holstein-Friesian X Local crossbred dairy cattle genotypes

Parameter	BWG1 (200 to below 300 kg/cow)	BWG2 (300 to below 400kg/cow)	BWG3 (400 to 570kg/cow)	F
MT	12.00 (4)	11.43±1.69 (7)	15.57±0.78 (35)	3.256
PM	13.50 (4)	13.71±1.69 (7)	18.00±0.78 (35)	4.155
LD	210.00 (4)	216.43±1.69 (7)	218.79±0.80 (33)	0.628

Note: MT-test day milk production in liter, PM-peak milk production per day in liter, LD-lactation period in one calving. F-F value in one way ANOVA during post-hoc analysis for DMRT

DISCUSSION

Correlation of MT with CP and GG

Strong positive correlation (0.794) between MT and CP reported. Similarly, positive correlation (0.453) between MT and GG was also documented. However, Body weight gain, milk yield and milk protein were increased as the ratio of concentrate feed was increased in the diet but milk fat was decreased (Sanh et al., 2002). Positive and significantly correlated traits might be suggested that, increase for one trait will affect the other positively. So, above discussion might be indicative that increase of CP and GG will affect the MT positively.

Effect of grades on MT, CPC and GGC

MT significantly varied among the grades and the highest MT was recorded for grade three and this was followed by grade two and one. However, present findings was similar to Bhuiyan et al. (2015), who reported 1473±35.3 liter of milk / cow/100days of lactation of the first calving daughters of Holstein Friesian crossbred bull in Bangladesh condition. On the other hand, the lowest amount of CPC was observed in grade three (9.99±0.70 BDT/liter), while the highest was noted in grade one. However, the highest amount of GGC was documented in grade one (0.82±2.00 kg/liter/cow) and the lowest were in grade three. Daily 7.14±0.72 kg concentrate feed supplied to the Friesian X local crossbred cow to produce 15.90±0.72 liter milk a day (Hossain et al., 2016). The above discussions might be suggested that grade three contributed more to increase MT and decrease CPC and GGC than grades two and one.

Effect of grades on PM and LD

PM and LD were affected by grades significantly and the highest amount of PM were recorded for grade three and the lowest were reported for grade one. On the contrary LD varied non-significantly between grades two and three but, the higher LD was recorded for grade three than two and one. On the other hand, Sarder et al. (2007), found longer lactation length of Holstein Friesian × Local crossbred cows (285±47days) than the present observation. The above discussions might be pointed that grade three contributed more to increase PM and LD than grades two and one.

Effect of age groups of cow on MT, PM and LD

Higher MT and PM were reported for age group 3 than age group 1 and 2. On the contrary, LD was higher in age group 2 than 1, though the same were higher in age group 3 than 2.However, Milk production performances for cows in 4th and more lactations were no longer better than that of their 3rd lactation and this might be due to more numbers of secretory cells after third lactations, were dieing as the cows becoming older and older (Epaphras et al., 2004).These non-significant variations among age groups might be suggested that age group 3 was better than 1 and age group 1 was better than 2 for MT and PM and higher MT and PM under age group 1 than 2, might be to the variation of body weight, genetic merit for milk production of experimental cows or it might be due to the variation of management. But apparently, age group 3 was better than 2 and age group 2 was better than 1 for LD.

Effect of body BWG on milk production performances under different grades

Grade one

All cattle were   with a body weight range like 200 kg to below 300 kg/cow under grade one crossbred cattle, mean lactation period in were lower than that of other grades and these lower body weights and LD might be due to the effect body weight and LD of dam (local cattle) of the experimental cows under this grade. Test day milk production and peak milk production per cow were close to the findings of Hossain et al. (2016), who observed, daily milk yield was higher in crossbred (15.90±0.72 liter/cow) than pabna (5.73±1.21 liter/cow) cattle at northern region of Bangladesh.

Grade two

In grade two, MT, PM and LD were affected by body weight groups significantly. The higher MT (18.75±2.62 liter/cow/day) and PM (20.75±2.62 liter/cow/day) were reported for BWG3 than BWG2 and BWG1. In a different study, Sarder *et al.* (2007) in Bangladesh found lower amount (11.63±2.90 liter/day/cow) of peak milk than the present findings for Friesian ×local crossbred cattle. But very interestingly, LD was reported higher in BWG2 and this was followed by BWG1 and BWG3 under the same grade. The above discussions might be indicative that BWG3 were superior for MT and PM but BWG2 was better for LD.

Grade three

In grade three, MT, PM and LD were not affected by body weight groups significantly but higher MT, PM and LD were reported for BWG3 under grade three. The above discussions might indicative that the higher the body weight the higher the MT, PM and LD could be found under grade three.

CONCLUSION

Increase of concentrate feed cost per cow (CP) and use of green grass per cow (GG) before test day milk production (GG) will increase the test day milk production (MT). Body weight group three under grade two was better for test day milk production per cow and peak milk production per cow (PM) but body weight group two was better for lactation period in days in one calving per cow (LD). Grades, body weight groups, CP and GG affected MT, PM, LD.

ACKNOWLEDGEMENT

We thankfully acknowledge the contribution of dairy cattle farmers at the villages of Sonaimuri, Senbagh, Suborno char and Noakhali sadar upazilas under the district of Noakhali in Bangladesh to support the research work through helping all the way to collect the data for this study.

REFERENCES

1. Banglapedia, 2014. Banglapedia, National Encyclopedia of Bangladesh. http://en.banglapedia.org/index.php?title=Cattle
2. BER, 2012. Bangladesh Economic Review (BER), Economic Division, Ministry of Finance. The Government of the People's Republic of Bangladesh, from http://www.mof.gov.bd
3. BER, 2013. Bangladesh Economic Review (BER), Economic Division, Ministry of Finance. The Government of the People's Republic of Bangladesh, from http://www.mof.gov.bd
4. BER, 2015. Bangladesh Economic Review (BER), Economic Division, Ministry of Finance. The Government of the People's Republic of Bangladesh, from http://www.mof.gov.bd
5. Bhuiyan AKFH, Rashid MM, Khan RA, Habib MA, Bhuiyan MSA, Faiz MA, 2015. Progeny tested bull production for dairy cattle development in Bangladesh. Bangladesh Journal of Animal Science, 44: 106-112.
6. Bhuiyan AKFH, 2011. Progeny Tested Seed Bull Production: Progress towards National Dairy Development. Keynote paper presented at the 3rd National Seminar of Breed Up-gradation through Progeny Test Project held on 15 June 2011 at BIAM, Dhaka.
7. Hossain MS, Islam F, Rashid MHO, Leena SA, Sarker SC, 2016. Productive and Reproductive Performances of Holstein Friesian ×Local Crossbred and Pabna × Pabna Cattle Genotypes. International Journal of Business, Social and Scientific Research, 4: 261-266.
8. Epaphras A, Karimuribo ED and Msellem SN, 2004. Effect of season and parity on lactation of crossbred Ayrshire cows reared under coastal tropical climate in Tanzania. Livestock Research for Rural Development, 16 (2).

9.   Hossain MS, Miah MY, Khandaker ZH, Islam F, 2015.Effect of different levels of matikalai (*Vignamungo*) hay supplementation to straw-based diets on feed intake, digestibility and growth rate of indigenous cattle.Livestock Research for Rural Development, 27 (2).

10. Huque KS, Dev GK, Jalil MA, 2011. High yielding dairy breed development in Bangladesh- limitations and opportunities. Paper presented in international workshop on "High yielding dairy breed development in Bangladesh" held at Bangladesh Livestock Research Institute, Dhaka from 28-29 September.

11. Sanh MV, Wiktorsson H, Ly LV, 2002. Effects of Natural Grass Forage to Concentrate Ratios and Feeding Principles on Milk Production and Performance of Crossbred Lactating Cows. Asian-Australasian Journal of Animal Science, 15: 650-657.

12. Sarder MJU, Rahman MM, Ahmed S., Sultana MR, Alam MM, Rashid MM, 2007. Consequence of dam genotypes on productive and reproductive performance of dairy cows under the rural condition in Bangladesh. Pakistan Journal of Biological Sciences 10: 3341-3349.

13. SPSS, 2005. Windows for version-14.0. Release on 27.10.2005. (Microsoft Corp. 1998). Trends SPSS Inc., Michigan Avenue, Chicago, IL. 19-182.

14. Tuan BQ, 2000. Effects of Protein and concentrate levels on rumen digestion and milk production of crossbed dairy cattle in Hanoi. PhD Thesis. University of Agriculture, Hanoi, Vietnam.

15. Usman T, Guo G, Suhail SM, Ahmed S, Qiaoxiang L, Qureshi MS and Wang Y,2012. Performance traits study of Holstein Friesian cattle under subtropical conditions. The Journal of Animal and Plant Sciences, 22: 92-95.

16

COMPARATIVE MARKET SUPPLY OF PROTEIN FROM LIVESTOCK AND FISH IN THE SELECTED URBAN AREAS OF RAJSHAHI DISTRICT IN BANGLADESH

Md. Mizanoor Rahman[1], Khan Shahidul Huque[2], Nani Gopal Das[2*] and Md. Yousuf Ali Khan[2]

[1]Department of Livestock Services (DLS), Farmgate, Dhaka, Bangladesh;
[2]Bangladesh Livestock Research Institute (BLRI), Savar, Dhaka-1341, Bangladesh

*Corresponding author: Nani Gopal Das; E-mail:nani.gd@hotmail.com

ARTICLE INFO

Key words

Livestock
Fish
Biomass supply
Protein supply
Rajshahi

ABSTRACT

The study was conducted with the objectives to determine the market availability of animal sourced foods (ASF) and fish, and their share in the supply of biomass and protein through visiting the wet markets of metropolitan and municipality areas of Rajshahi district, Bangladesh. A preset questionnaire was used for recording the biomass weight of different ASF and fish in every four days interval in March, 2016. It was found that the supply of ASF (beef, chevon, chicken and egg) and fish in the metropolitan markets (80.20 and 35.89 t/d, respectively) was significantly higher (P<0.05) than any municipality wet market in the district (7.66 and 3.03 t/d, respectively). The market supply of biomass and its protein value of ASF were 3.64 and 4.33 times higher than fish. The chicken shared the highest amount of protein (28.19 %) followed by fish (26.8%), beef (26.21%), eggs (11.46%) and chevon (7.34%) during the study period. However, this initial work does not include milk, and the wet market of ASF and fish may have seasonal variations which needs to be explored through further research. In addition to them, socioeconomic status of consumers and regional variations are important which needs to be studied for addressing resource base safe food production help the strategic reduction of food insecurity in the country by 2030.

INTRODUCTION

Animal sourced foods (ASF) include food items come from an animal source such as, meat, milk, eggs and their products. Both ASF and fish are the sources of six critical micronutrients (vitamin A, vitamin B12, riboflavin, calcium, iron and zinc) and essential amino acids required for human health. An inadequate intake of these nutrients is associated with negative health impacts (Murphy and Allen, 2003). The generally accepted daily dietary allowance of total protein is 0.8g/kg live weight (Shane and Mann, 2006) of human and thus, an adult person of 60 kg requires 48g protein daily, of which an average of 24g (50%) animal protein may come from ASF and fish sources.

The per caput ASF (milk, meat and egg) production in 2014-15 was 121.8 ml, 102.4 g and 9.61 g, respectively (BER, 2016), and they may supply daily about 26.06 g protein/head, while the production of fish was 64.4 g that yields 10.94 g protein (considering average 17% protein of different fish of Bangladesh, Bogard *et al.*, 2015). Thus, ASF & Fish sharing 70.4% and 29.6%, respectively, may daily supply 37.0 g/head of animal protein, almost 1.54 times more than the animal protein requirement (24.0 g) for an average person of the country. The Household Income and Expenditure Survey, 2010 (HIES- 2010) in BBS, (2014), on the other hand, reported per caput protein intake of 6.76g and 9.70g in 2010 from ASF and fish and they may stand to 7.94 g and 12.85 g in 2015 sharing 38.0% and 62.0%, respectively, of the two sources if an extrapolation is made considering the average annual growth rate of livestock & fisheries sub-sectors during the five years period (BER, 2015). This does not conform with the production of ASF and fish and their shares (70.4% vs 29.6%) in total animal protein production described above, and opposes the claim of about 60% of the daily animal protein supply of a person comes from fish made by the Department of Fisheries (MoFL, 2016).

This divergent and anecdotal evidence of the two different documents of the public sector requires a thorough field study by the concerned authority on the production and supply of ASF and fish to different classes of consumers considering seasonal and regional variations in the country. Recently, the United Nations Statistical Commission also requested all member countries to collect and analyze their own experience-based food insecurity data to achieve indicator 2.1.2 of Sustainable Development Goal (SDG) 2.1 and use them for national, regional and global reporting (FAO 2016). Considering the above data oscillation and to gain experience-based protein shares of the ASF and fish, the present initial field research was undertaken with a simple objective of estimating the market supply of ASF and fish in a certain season of a year in a selected area of the country.

MATERIALS AND METHODS

Study area

The study was conducted at ten municipality areas of Rajshahi district, a High Ganges River Floodplain agro-ecological zone in the north-western area of the country (Figure 1). Three wet markets of Boalia of Rajshahi metropolitan area and of each municipality area (Tanore, Mohanpur, Paba, Godagari, Durgapur, Puthia, Bagmara, Charghat and Bagha) of Rajshahi district were randomly selected for this study. The total area and household number of the metropolitan and municipality areas according to BBS (2014) are shown in Table 1.

Figure 1. Metropolitan and municipalities under the study

Data collection

The total wet markets and the number of traders interviewed in each municipality and the metropolitan area was recorded and reported accordingly (Table 2). The amount of ASF and fish marketed in every four days interval in randomly selected three wet markets were weighed and recorded through visiting and recording the weight of ASF and fish biomass. Under the direct supervision of district livestock office, Rajshahi a preset questionnaire was used for recording the biomass weight of different ASF and fish and responses of the wholesalers and retailers; seven visits were made during the total period of 31 days in March, 2016. The daily average supply of ASF and fish in each municipality was calculated by multiplying the daily average supply in selected markets by the total number of wet markets.

Table 1. Area and number of households in different municipalities of Rajshahi district

Name of municipalities	Area (sq km)	Households
Rajshahi metropolitan (Boalia)	38.11	42602
Mohanpur	9.25	2590
Paba (Katakhali and Noahata)	49.59	20998
Godagari	14.29	8008
Puthia	13.50	5188
Durgapur	24.83	7109
Tanore	27.22	7976
Charghat	18.72	9105
Baghmara (Bhawaniganj and Taherpur)	24.18	7231
Bagha	11.69	7044

Calculation of protein supply

The daily protein supply of beef and chevon biomass was calculated according to the following equation: Protein (kg) = biomass of beef/chevon (kg) × meat (%) of total biomass × average protein (%) (Roy *et. al.*, 2013; Sumarmono *et. al.*, 2002); while that of chicken was calculated according to the equation: Protein (kg) = supply of chicken biomass (kg) × dressed weight (%) × meat (%) of total biomass × protein (%) (Połtowicz and Doktor, 2011). The supply of protein from eggs was calculated by multiplying biomass with the protein content of whole egg according to Bashir *et al.* (2015). In case of fish the biomass was converted into edible portion according to Akther (2015) and then multiplied by the average protein content of fish in Bangladesh according to Bogard *et al.* (2015).

Data analysis

Any significant difference in the supply of biomass and protein of different ASFs and the fish in addition to their share between the metropolitan and municipality areas were tested statistically using Student T-test. A computer package program of SPSS-11.5 was used for data compilation and analyses.

RESULTS AND DISCUSSION

The supply of ASF and fish biomass

The daily market supply of ASF and fish biomass in a wet market of the Rajshahi metropolitan and of municipality areas are presented in Table 3. It was seen that the market availability of all ASF and fish biomass was significantly (P<0.01) higher in Rajshahi (Boalia) metropolitan market compared to municipalities of the district. The daily supply of beef in a wet market of Rajshahi metropolitan area was 16.49 t compared to 3.49 t in a municipality area, and it was 4.72 times higher. Similar to beef, the daily market supply of chevon was 7.01 t and 0.59 t, respectively. The higher supply of beef and chevon resulted in a total daily red meat supply of 23.05 t in the metropolitan market and 4.08 t in a municipality market, and the difference between the market sources was significant (p<0.05). Including 40.79 t daily supply of poultry meat (white meat) the total daily meat supply in a wet market of Rajshahi metropolitan area was 64.29 t and the supply of both poultry meat (2.10 t) and the total meat (6.18 t) was significantly (p<0.05) lower in the municipality markets.

The market supply of eggs at Rajshahi (Boalia) metropolitan and municipality area were 15.91 and 1.41 t/d, respectively, and they resulted in a total market ASF supply of 80.20 t and 7.66 t/d, respectively. The difference between the two sources were significant (p<0.01). The daily supply of fish, on the other hand, was 35.89 t and 3.03 t, respectively in Rajshahi (Boalia) metropolitan and other municipality markets.

An increased number of consumers due to a higher concentration of households in a unit metropolitan area (average 1118 households) coupled with a higher income (HIES - 2010 in BBS, 2014) compared to that of municipality area (280 to 602 households, Table1) increased the number of consumers for ASF and fish. A positive income elasticity of demand of food of livestock origins (Gandhi and Zhou, 2010) and a higher number of consumers may have increased the market demand of ASF and fish in the former area. This provided an increased opportunity of ASF and fish trading in the metropolitan area. The relationship between the number of traders and supply of meat and fish in the market shows that the market biomass supply of meat or fish increased linearly (r=0.84, P<0.01, Figure 2) with the increase of trader number (Figure 2). It was calculated that the daily market supply of meat by a trader of the studied areas, irrespective of red or white meat, was about 4.05 kg, and that of fish was about 4.90 kg (r=0.84, P<0.01, Figure 2 and 3).

Table 2. Total number of different wholesalers/retailers interviewed

Name of municipalities	Number of wet markets	Number of whole sellers & retailers					
		Beef	Chevon	Chicken	Egg	Total ASF	Fish
Rajshahi metropolitan (Boalia)	20	51	28	59	51	189	153
Mohanpur	10	9	2	37	7	55	50
Paba (Katakhali and Noahata)	10	21	8	10	14	53	56
Godagari	11	10	0	5	22	37	22
Puthia	10	17	6	16	26	65	88
Durgapur	6	11	4	16	19	50	45
Tanore	3	11	5	14	9	39	9
Charghat	9	15	8	8	19	50	18
Bagmara (Bhawaniganj and Taherpur)	11	13	14	13	23	63	36
Bagha	8	32	42	17	32	123	40
Total traders	98	190	117	195	222	724	517

Table 3. The average market supply of ASF and fish biomass between Rajshahi metropolitan (Boalia) and municipality areas (t/d)

Food sources	Market biomass supply (t/d)		SE	P- values
	Metropolitan (n = 3)	Municipality (n = 27)		
Beef	16.49	3.49	2.90	<0.01
Chevon	7.01	0.59	2.62	<0.01
Total red meat	23.50	4.08	4.49	<0.01
Poultry meat (White meat)	40.79	2.10	9.36	<0.01
Total meat supply	64.29	6.18	13.19	<0.01
Eggs	15.91	1.48	6.30	<0.01
ASF	80.20	7.66	18.70	<0.01
Fish	35.89	3.03	7.43	<0.01

SE=standard error; P>0.05, not significant; ASF, beef, chevon, chicken and egg

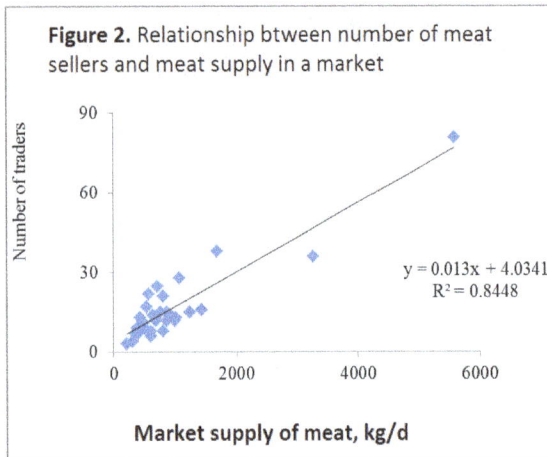

Figure 2. Relationship btween number of meat sellers and meat supply in a market

$y = 0.013x + 4.0341$
$R^2 = 0.8448$

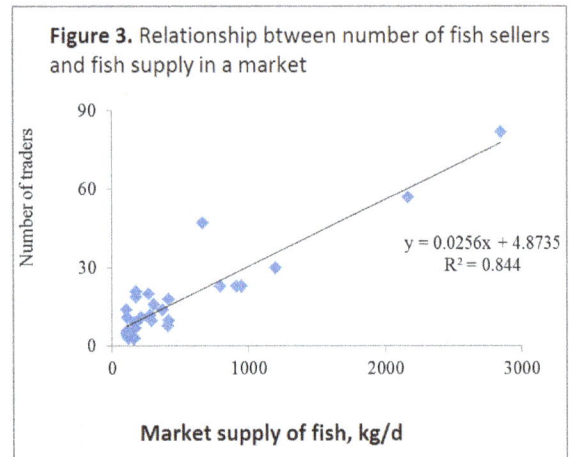

Figure 3. Relationship btween number of fish sellers and fish supply in a market

$y = 0.0256x + 4.8735$
$R^2 = 0.844$

The protein value of ASF and fish biomass

The supply of protein of ASF and fish in the market of Rajshahi metropolitan and municipality area is presented in Table 4. Similar to biomass supply, the market supply of protein of all ASF and fish sources was significantly higher in Rajshahi (Boalia) metropolitan areas compared to that of the municipality area of the district (P<0.01). The protein value of beef and chevon marketed daily at the wet market of Rajshahi (Boalia) metropolitan was 2.62 t and 1.22 t, respectively, compared to only 0.55 t and 0.10 t of the municipality area. The protein content of the total red meat supply was calculated to be 3.83 t and 0.66 t, respectively. The protein value of poultry meat in Rajshahi (Boalia) metropolitan wet market was 5.53 t and it was significantly (p<0.01) higher than that (0.29 t) of the wet market in the municipality area, and the difference in protein value of market poultry between the two area was about 19.07 times. The protein value of the total daily meat supply was 9.37 t and 0.94 t, respectively, in the wet market of Rajshahi (Boalia) metropolitan and the municipality area. The protein content of eggs supplied in the markets of Rajshahi (Boalia) metropolitan per day was 1.78 t compared to 0.17 t of that supplied daily in the markets of municipality areas of the district. The total protein value of the ASF daily supplied to Rajshahi (Boalia) metropolitan wet market was 11.15 t and that of the municipality markets was 1.11 t. At the same time the protein value of fish marketed daily in Rajshahi (Boalia) metropolitan wet market was 4.41 t compared to 0.37 t in the wet market of municipality area.

Table 4. Protein supply from ASF and fish at Rajshahi (Boalia) metropolitan and other municipality markets (t/d)

Food Sources	Protein supply (t/d)		SE	P- values
	Metropolitan (n = 3)	Municipality (n = 27)		
Beef	2.62	0.55	0.46	<0.01
Chevon	1.22	0.10	0.46	<0.01
Total red meat	3.83	0.66	0.74	<0.01
Poultry meat (white meat)	5.53	0.29	1.27	<0.01
Total meat supply	9.37	0.94	1.92	<0.01
Eggs	1.78	0.17	0.70	<0.01
ASF	11.15	1.11	2.53	<0.01
Fish	4.41	0.37	0.91	<0.01

SE= standard error; P>0.05, not significant; ASF, beef, chevon, chicken and egg

Table 5. Share of different ASF and fish in the supply of biomass and protein

Food sources	Ratio of ASF components and fish supply				
	Metropolitan (n =3)	Municipality (n = 27)	SE	P- values	Total (n = 30)
Biomass components					
Total meat: fish	1.88	3.32	2.35	0.324	3.18
Red meat: fish	0.86	2.12	1.73	0.242	1.99
Chicken: fish	1.02	1.20	0.88	0.736	1.18
ASF: fish	2.28	3.79	2.45	0.321	3.64
Protein components					
Total meat: fish	2.26	4.09	2.93	0.313	3.91
Red meat: fish	1.13	2.76	2.26	0.246	2.60
Chicken: fish	1.12	1.33	0.97	0.734	1.31
ASF: fish	2.62	4.52	3.02	0.311	4.33

SE=standard error; P>0.05, not significant; n, number of replication; ASF, beef, chevon, chicken and egg

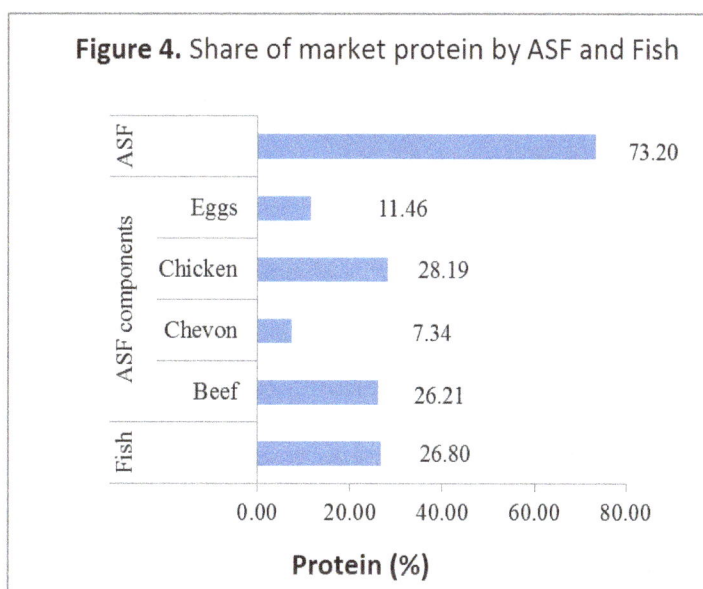

Figure 4. Share of market protein by ASF and Fish

The comparative share of biomass and protein of different ASF and fish

The ratio of market biomass and protein supply from ASF and fish are presented in Table 5. It was found that the share of ASF components and fish biomass marketed in the metropolitan and municipality area and their protein values did not differ significantly (P>0.05). The total meat supply in metropolitan and municipality areas of Rajshahi was 1.88 and 3.32 times higher than fish. Compared to the market supply of fish biomass, the supply of red meat and chicken in both metropolitan and municipality area was 0.86 and 1.02 times in the metropolitan area and 2.12 and 1.20 times in the municipality area. However, the market supply of ASF in former markets was 2.28 times compared to 3.79 times in the latter. The average biomass supply of total meat, red meat, chicken and AFS compared to that of fish, irrespective of areas, was 3.18, 1.99, 1.18 and 3.64 times higher, respectively.

The supply of protein of total meat, red meat, chicken and AFS was 2.26, 1.13, 1.12 and 2.62 times, respectively, and 4.09, 2.76, 1.33 and 4.52 times, respectively, higher than that of fish in metropolitan and municipality area, respectively; and their average, irrespective of metropolitan and municipality, was 3.91, 2.60, 1.31 and 4.33, respectively in the study area. The global average per caput meat (both red and white meats) consumption is found to be always higher (43.2 Kg in 2015) than fish (20.3 Kg) (FAO, 2016).

The development of database on the consumption of per caput meat of different animals, not considered here, is much more important, and it has to be species, region and season wise in one hand and on the other consumer categories have to be taken into consideration. Moreover, milk and value added products of different meat and fish should also be taken into consideration. Value additions, in one way, support the widening of gross domestic product basket, and on the other, it supports safe food production from field to forks. The highest share of protein value of fish and different AFS marketed during the study period was found by chicken (28.19 %) followed by fish (26.8%), beef (26.21%), eggs (11.46%) and chevon (7.34%, Figure 4), and it resulted in a total ASF share of 73.20% in Rajshahi district.

CONCLUSION

It may be concluded that the daily market supply of total meat and ASF (64.29 t and 80.2 t) or their protein value (9.37 t and 11.15 t) were higher than that of fish (35.89 t) or its protein (4.41 t), irrespective of areas. The share of protein of chicken marketed daily was the highest (28.19 %) followed by fish (26.8 %), beef (26.21 %), eggs (11.46 %) and chevon (7.34 %). However, considering the importance of ASF and fish for addressing the food insecurity of the people, further research may be conducted on the production, marketing, value additions and their per caput consumption according to season, region and socioeconomic conditions. This will enable formulation of livestock and fisheries resource based strategic development plans in the country.

CONFLICT OF INTEREST

The author declares that there is no conflict of interests regarding the publication of this paper.

REFERENCES

1. Akther S, 2015. Flesh yield of small indigenous fishes from the river Padma, Rajshahi, Bangladesh. Bangladesh Journal of Zoology, 43: 141-144.
2. Bashir L, PC Ossai, OK Shittu, AN Abubakar and T Caleb, 2015. Comparison of the Nutritional Value of Egg Yolk and Egg Albumin from Domestic Chicken, Guinea Fowl and Hybrid Chicken. American Journal of Experimental Agriculture 6: 310-316.
3. Bangladesh Bureau of Statistics (BBS), 2014. Statistical yearbook Bangladesh, Ministry of Planning, Government of the People's Republic of Bangladesh, Dhaka.
4. Bangladesh Economic Review (BER), 2015. Economic advisor's wing. Finance Division, Ministry of Finance. Government of the People's Republic of Bangladesh.
5. Bangladesh Economic Review (BER), 2016. Economic advisor's wing. Finance Division, Ministry of Finance. Government of the People's Republic of Bangladesh.
6. Bogard JR,SH Thilsted, GC Marks, MA Wahab, MAR Hossain, J Jakobsen and J Stangoulis, 2015. Nutrient composition of important fish species in Bangladesh and potential contribution to recommended nutrient intakes. Journal of Food Composition and Analysis, 42:120–133.
7. FAO, 2016. Food and Agriculture Organization (FAO) of the United Nations. Food Outlook; Biannual report on Global food markets. http://www.fao.org/3/a-i6198e.pdf
8. Gandhi, VP and ZY Zhou, 2010. Rising demand for livestock products in India: Nature, patterns and implications, Australasian Agribusiness Review. 18:103-135.
9. MoFL, 2016. Annual report- 20015-16. Ministry of Fisheries and Livestock. The Peoples` Republic of Bangladesh. p: 20.
10. Murphy, SP and LH Allen, 2003. Nutritional importance of animal sourced food. Journal of Nutrition, 133: 3932-3935.

11. Połtowicz, K and J Doktor, 2011.Effect of free-range raising on performance, carcass attributes and meat quality of broiler chickens. Institute of Genetics and Animal Breeding, Jastrzębiec, Poland. Animal Science Papers and Reports, 29: 139-149.

12. Roy BK, KS Huque, NR Sarker, SMJ Hossain, MS Rana and MN Munsi, 2013. Study on growth and meat quality of native and Brahman crossbred bulls at different ages. Proceedings of the annual research review workshop- 2012-13, p. 81-116.

13. Shane B and N Mann, 2006. A Review of issues of dietary protein intake in humans. International Journal of Sport Nutrition and Exercise Metabolism, 16:129–152.

14. Sumarmono J, NMW Pratiwi, PJ Murray and DG Taylor, 2002. Yield and carcass characteristics of improved boer and Australian Feral goats slaughtered at 30 kg live weight. Animal Production in Australia, 24: 233-236.

PROBLEMS AND PROSPECTS OF TURKEY (*Meleagris gallopavo*) PRODUCTION IN BANGLADESH

Mohammad Asaduzzaman[1], Ummay Salma[2], Hussein Suleiman Ali[2], Md. Abdul Hamid[2] and Abdul Gaffar Miah[1*]

[1]Department of Genetics and Animal Breeding and, [2]Department of Animal Science and Nutrition, Faculty of Veterinary and Animal Science, Hajee Mohammad Danesh Science and Technology University, Dinajpur, Bangladesh

*Corresponding author: Abdul Gaffar Miah, E-mail: agmiah2009@gmail.com

ARTICLE INFO	ABSTRACT

The study investigated the production status, problems and prospects of turkey production in Bangladesh following survey and multistage sampling procedure. Average flock size, weight of a tom and hen were 15.34±2.38, 6.58±0.15 and 2.39±0.06 kg, respectively. Commercial, homemade, and both homemade and commercial feed were used by 21.74, 30.43 and 47.83% farmers, respectively. Both tom and hen attained puberty at 7.22±0.06 months, a hen laid 69.46±0.78 eggs *per annum* and weight of each egg was 66.13±0.63 g. Fertility and hatchability of eggs were 50±3 and 32±1%, respectively. Male and female ratio maintained 1:4.60±0.17. Main reasons of lower hatchability were low egg fertility, faulty incubation, and both low egg fertility and faulty incubation as per 50.0, 21.7 and 28.3% farmers, respectively. None of the farmers used artificial insemination (AI) except natural breeding. Main advantages of turkey rearing over other poultry species were low disease, high market price, low feeding cost and low mortality according to 41.3, 28.3, 17.4 and 13.1% farmers, respectively. While 36.9% farmers had encountered disease, 80.4% had not used vaccine. An egg, a poult and an adult turkey were sold at BDT 76.2±1.79, 838.5±22.8 and 2587.2±74.8, respectively. In fact, turkey production is still at primitive stage which is characterized by poor housing, feeding, breeding and healthcare practices, so vigorous public extension service, training, research and marketing strategies are immediately needed to improve this sector in Bangladesh.

Key words

Turkey
Fertility
Hatchability
Production
Marketing

INTRODUCTION

Bangladesh is a small country with a large population about 160 million, situated between 88°10' and 92°41' East longitudes and between 20°34' and 26°38' North latitudes in south Asia with flat land area (147,570 sq.km). Traditional backyard poultry keeping has been practiced in this country since time immemorial. Besbes (2009) reported that the worldwide poultry sector consists of chickens (63%), ducks (11%), geese (9%), turkeys (5%), pigeons (3%) and guinea fowls (3%). From the last decade, demand for poultry products has been increased rapidly in Bangladesh, and propelled by rising levels of income, population and urbanization. Experience shows that climate of Bangladesh is convenient to rear different poultry species. Poultry meat alone contributes 37% of the total meat production in Bangladesh (Begum et al., 2011).

Poultry transform feed into animal protein very rapidly. Poultry consumption in developing countries is projected to grow at 3.4% *per annum* to 2030, followed by beef at 2.2% and ovine meat at 2.1%, and in the world as a whole, poultry consumption is projected to grow at 2.5% *per annum* to 2030, with other meats growing at 1.7% or less (FAO, 2007). The environmental impact of poultry production is a continuing challenge and it is predicted that global consumption of poultry meat will increase between 2000 and 2030 at an average annual rate of 2.51% (Fiala, 2008).

In fact, poultry keeping is an integral part of the rural household that provides family income for the small, marginal and landless poor. The farmers who cannot afford to rear cattle and goat can easily rear poultry. However, among the livestock sector, the poultry industry (specially, commercial broiler and layer) is in the line to be destroyed due to severity of avian influenza (bird flu). Thus, it is crying need to search the alternative protein source to meet up the increasing demand. In order to maximize food production and meet protein requirements in developing countries, variable options need to be explored and evaluated (Owen et al., 2008). Turkey meat may be a one of the best options for alternative protein source in Bangladesh. Turkey production is an important and highly profitable agricultural industry with a rising global demand for its products (Yakubu et al., 2013), and they are adaptable to wide range of climatic conditions (Ogundipe and Dafwang, 1980). Karki (2005) stated that consumption of turkeys and broilers as white meat was rising worldwide and a similar trend also existed in developing countries. In the whole world, total production of turkey meat was 5.6 million ton in 2012, which was higher than 5.1 million ton in 2003, a decade earlier (FAOSTAT, 2012). Turkey is an excellent insect forager and most crops that are troubled by insect population including vegetables are candidates for insect control by turkeys (Grimes et al., 2007). Turkey thrives better under arid conditions, tolerates heat better, ranges farther and has higher quality meat (Yakubu et al., 2013). But turkey production has not been fully exploited in Bangladesh including other developing countries despite its huge potential over other poultry species.

In fact, turkey is a newly introduced poultry species in Bangladesh. Farmers are rearing turkey as an ornamental bird with a limited extent without having prior experience. Mainly interested farmers started turkey farming by importing day-old turkey chicks (Poult) from neighboring country, India. Its popularity is increasing gradually because of gamey flavor of meat with lower fat content. So, it may have high potential for production and marketing in Bangladesh. However, there is scanty study conducted previously regarding turkey production in Bangladesh. Therefore, the study has been undertaken to investigate the present status and production system of turkey and turkey farmers, and the problems and prospects of turkey production and marketing in Bangladesh.

MATERIALS AND MEHTODS

Study site and duration of the experiment

The study area was different districts of seven (7) divisions of Bangladesh. Famers of Dhaka, Gazipur, Rangpur, Dinajpur, Nilphamari, Bogra, Narshighdi, Mymensingh, Natore, Naogaon, Pabna, Sirajgonj, Chittagong, Sylhet, Khulna, Satkhira, Barisal, Kushtia, Rajbari and Noakhali districts were interviewed (Figure 1). The study period was from February to June 2016.

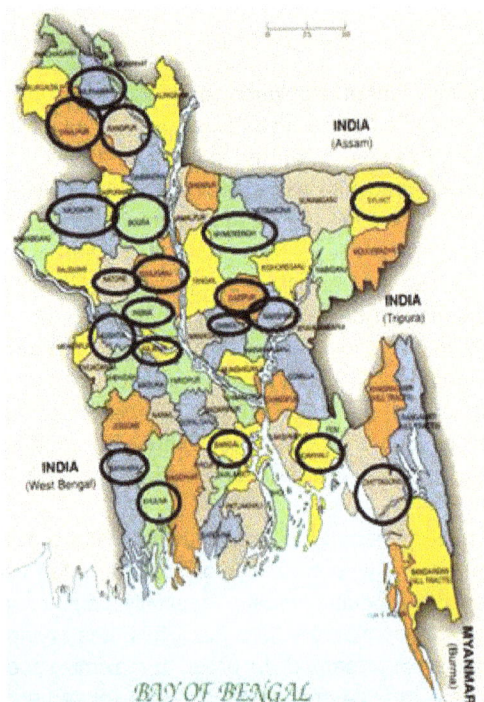

Figure 1. Black circle on the map of Bangladesh showing the study districts (*source: www.lged.gov.bd*)

Experimental design

Exploratory research design was followed to conduct the study. The methods-survey, review of secondary data, interview, observation and Focus Group Discussion (FGD) were conducted taking representative sample from all over Bangladesh. The questionnaire was carefully designed keeping in mind the objectives. The questionnaire contained both open and closed forms of questions. Most easy, simple and direct questions were asked to obtain information. The questionnaire was pre-tested with three (3) farmers for judging suitability for the farmers. After having feedback from field test, necessary modifications were done and the questionnaire was finalized for data collection.

Sampling Technique and Sample Size

A multistage sampling procedure was followed. Purposive sampling procedure was followed to select districts from seven divisions giving focus on concentration of turkey farms. There was no list of turkey farmers in the hand of any government and non-government agencies because of its newness in Bangladesh. So keeping in view the objectives of the study, a list of 56 turkey farmers was prepared from all over Bangladesh via personal communication, Facebook, bikroy.com and other sources. Simple random sampling technique was used to select 49 turkey farmers. The sample size of the respondents was determined by using proportion sample formula: the Slovin's formula (Adanza, 2006).The formula is presented below:

$$n = \frac{N}{1+N(e)^2}$$

Where;

n is the sample size sought;

N is the research population,

e is the level of confidence (taken as 95%).

The sample size (n) for this study calculated using the formula was:

$$n = \frac{56}{1+56(0.05)^2} = 49.$$

Therefore, primary data were collected from 49 respondent farmers selected from different divisions of Bangladesh.

Data collection

Direct observation, interview and farm record analysis methods were applied during collecting data for the study. Primary data were collected from turkey farmers were on farmers' personal information (age and education level), housing, feeding, breeding, management, disease, marketing, problems and prospects. Some parameters like flock size, number of egg production, weight of egg, male and female ratio etc. were taken. The sources of secondary data were review of literature from official documents, Journals, libraries, research institutes, internet etc. Participatory Rural Appraisal (PRA) tools like Focus Group Discussion (FGD), seasonality analysis of disease and market etc. were also used in relevant cases to collect and verify data. The researcher performed all the interviews to ensure consistency in data quality.

Statistical analysis

Collected data were complied, tabulated and analyzed. Qualitative data were converted into quantitative forms by means of suitable score whenever needed and the local units were converted into standard unit scales. Data were analyzed using the Statistical Package for Social Sciences (SPSS) program package (SPSS, 2013). Simple descriptive statistics such as frequency distributions, percentages, mean and standard error of mean (SEM) were applied to illustrate the results.

Problems encountered during the study

Travelling to remote village was a big problem. Sometimes paraphrasing of scientific terms took time. Some of the farmers hesitated to answer questions regarding giving information on source of turkey, profitability and hatching technique. Sometimes farmers were not available on the scheduled time because of family and social obligations and other business reasons.

Limitations of the study

There was limitation of transport to meet those farmers who were living in remote areas. Most of the farmers used to not keep record of farming activities properly. So, farmers provided information recalling their memories. For this reason, in some cases value judgment was applied to have necessary data.

RESULTS

General farming management status and practices

Demography of farmers

To understand demographic and socioeconomic context of existing turkey farmers' data on age, education, sex, access to technical support and prior experience of other farming were collected. The average age of the respondent farmers was 40.5±1.38 years. Ownership of 98% turkey farms was belonged to the male farmers. Duration of turkey farming of the respondents was 20.2±1.56 months. The study showed that 4.34, 4.35, 26.09, 47.83 and 17.39% farmers obtained educational qualification <SSC, SSC, Bachelor and Masters Degree, respectively (Figure 2).

Purpose of turkey rearing

Turkey rearing is a new farming activity in Bangladesh. The study showed that 34.78, 19.57 and 45.65% respondent farmers were rearing turkey for ornamental, both egg and meat, and both meat and ornamental purposes, respectively.

Source of receiving technical support

The farmers seek technical support from different sources. The study showed that 10.88, 10.87, 28.26 and 50.00% farmers took technical support from Department of Livestock Services (DLS), both internet and DLS, internet and other farmers, and other farmers, respectively (Figure 3).

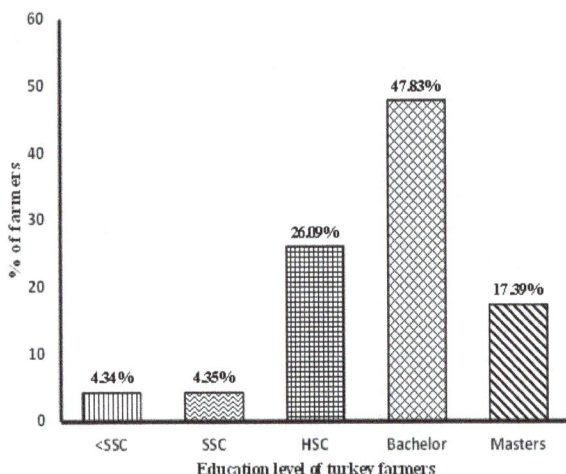

Figure 2. Education level of interviewed turkey farmers of Bangladesh

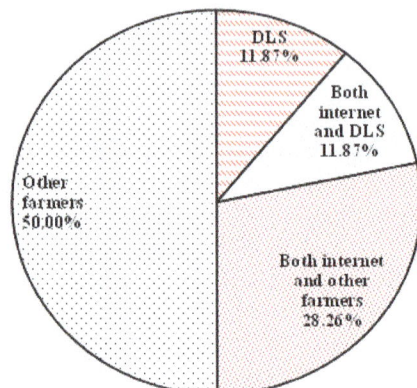

Figure 3. Source of receiving technical support for turkey farmers

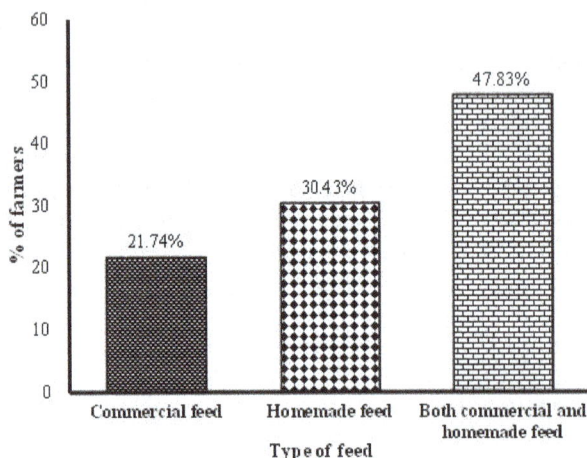

Figure 4. Type of feed used by farmers for feeding turkey

Figure 5. Wastage of feed in traditional turkey rearing system

Flock structure

The results obtained from the study on flock structure are presented in Table 1. It showed that average flock size of turkey was 15.34±2.38. The number of male (Tom) and female turkey (hen) ownership were 5.43±1.02 and 8.65±1.51, respectively. On the other hand average age of male and female turkeys found was 9.79±0.32 and 8.64±0.26 months, respectively. Average weight found were 6.58±0.15 and 2.95±0.06 kg for male and female turkey, respectively. Most of the farmers (88.86%) were raising both white and black color turkey birds while 9.25 and 1.89% were raising only white and black variety, respectively.

Housing and management

Results on turkey housing showed that 3.41, 64.92 and 31.67% farmers were raising turkey in free range, semi-intensive and intensive system, respectively. While 51.42% farmers informed that they took extra care during hot period for comfort of turkey, 48.58% did not take any extra care. On the other hand, 27.55% farmers took additional care during winter season while 72.45% did not take additional care. During winter and summer season farmers followed brooding period for 20.34±0.73 and 7.86±0.40 days, respectively (Table 1). It was found that while 16.25% farmers followed lighting procedure for breeder turkey, 83.75% had not followed. Most of the farmers did not use scientifically constructed nest box to facilitate laying of eggs by hen.

Feeding

The study showed that 21.74, 30.43 and 47.83% farmers used commercial, homemade, and both homemade and commercial feed, respectively for feeding their turkey (Figure 4). None of the interviewed turkey farmers calculated feed efficiency (FE) and wastage of feed found happened in many farms due to lack of using proper feeding methods (Figure 5).

Table 1. Average data on general farming management of turkey farming in Bangladesh

Parameters	Mean± SEM
Age of farmers (year)	40.54±1.38
Experience of turkey farming (months)	20.19±1.56
Flock size (number)	15.34±2.38
Number of male turkey	5.43±1.02
Number of female turkey	8.65±1.51
Age of Tom (month)	9.79±0.32
Age of Hen (month)	8.64±0.26
Length of brooding in winter (days)	20.34±0.73
Length of brooding in summer (days)	7.86±0.40
Price of adult turkey (BDT)	2587±74.7
Price of egg (BDT)	76.15±1.79
Price of one month old poult (BDT)	838±22.8

Table 2. Average productive and reproductive performance of turkey in Bangladesh

Parameters	Mean± SEM
Weight of adult Tom (kg)	6.58±0.15
Weight of adult Hen (kg)	2.39±0.06
Weight of egg (g)	66.13±0.63
Fertility percent of turkey egg (%)	50.00±3.00
Hatching percent of turkey egg (%)	32.00±1.00
Egg production/hen/year (No.)	69.46±0.78
Number of clutch in a year (No.)	2.3±0.01
Egg production in a clutch (No.)	25±0.80
Duration of a clutch (month)	2.2±0.3
Ratio of male to female ($\male$:$\female$)	1: 4.60±0.17

Health Management

The study showed that while 36.96% farmers had encountered diseases like New Castle disease, Fowl cholera, Fowl pox, Mycoplasmosis etc., 63.04% had not experienced any disease. Similarly, while 19.57% farmers had used vaccine, 80.43% had not used any vaccine (Figure 6).

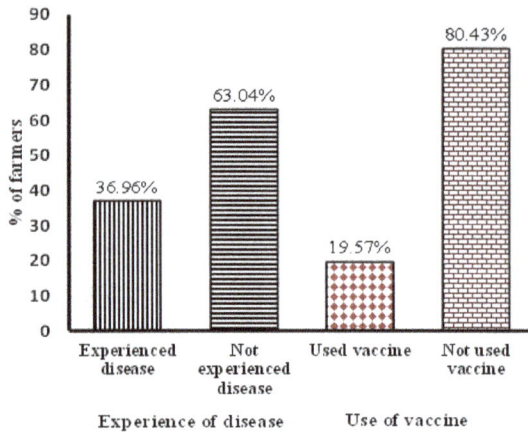

Figure 6. Farmers' experience on turkey disease and use of vaccine

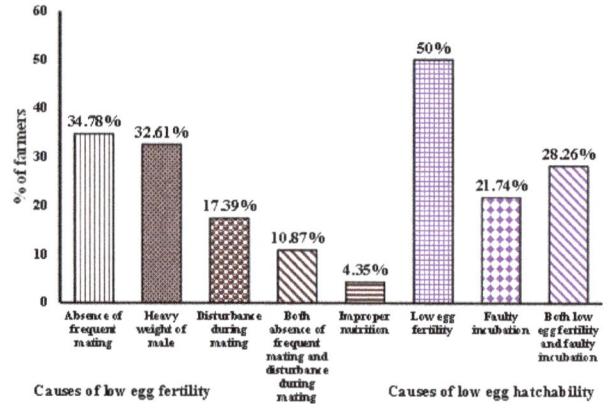

Figure 7. Farmers' perception on causes of low egg fertility and hatchability

Marketing

Results showed that farmers sold an egg, a poult and an adult male/female turkey at the rate of BDT 76.15±1.79, 838±22.8 and 2587±74.8, respectively (Table 1). Farmers did not keep record for which purpose the customers purchased turkey. Usually, customers who intended to farming, purchased turkey in pair i.e. one male and one female.

Productive and reproductive performance

Productive and reproductive performances of turkey are presented in Table 2. Average weight of the tom and hen found 6.58±0.15 and 2.39±0.06 kg, respectively. Farmers' experiences revealed that both tom and hen attained puberty at the same age and it was 7.22±0.06 months. A hen laid on an average 69.46±0.78 eggs per annum and weight of each egg was 66.13±0.63 g.

Male-female ratio and farmers experience on fertility

Male and female ratio maintained by the interviewed farmers was 1:4.60±0.17. Average fertility of turkey egg was experienced by the respondent farmers was 50.00±3.00%. In case of low fertility 34.78, 32.61, 17.39, 10.87, and 4.35% farmers identified the main reason as absence of frequent mating, heavy weight of male, disturbance during mating, both absence of frequent mating and disturbance during mating and improper nutrition in diet, respectively (Figure 7 and 8).

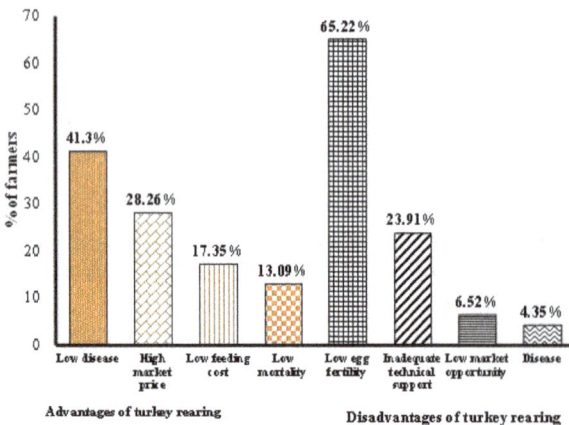

Figure 10. Farmers' perception on advantage and disadvantage of turkey

Figure 11. Improper brooding of turkey poult with chick and duckling by farmers

Figure 12. Foraging of turkey flock at farmer's garden

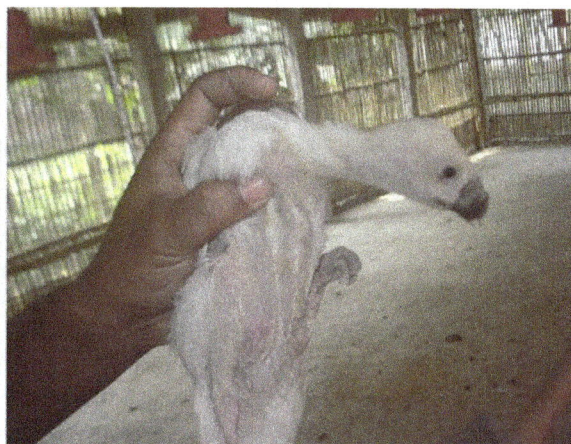

Figure 13. Severe attack of fowl pox at poult stage of turkey

Farmers experience of egg hatchability

Farmers experienced 32.00±1.00% hatchability of eggs which indicated lower fertility and not viable from business point of view. With this regard, 50.00, 21.74 and 28.26% farmers opined that the main reason of lower hatchability were low egg fertility, faulty incubation, and both low egg fertility and faulty incubation, respectively (Figure 7). Results on using hatching medium of turkey eggs showed that 10.8, 18.9, 37.8 and 27.0% farmers hatched their eggs using turkey hen, chicken hen, both turkey and chicken hen, and incubator, respectively (Figure 9).

Breeding methods used by turkey farmers

All the interviewed farmers followed natural breeding for reproduction of turkey. None of the farmers used artificial insemination (AI) as an assisted reproductive technique for turkey breeding.

Clutch size

Farmers experienced 2.3±0.01 clutch for turkey hen in a year. Average egg production in each clutch was 23.37±0.80. Duration of each clutch was 2.2±0.3 month.

Farmers' perception on problems of turkey farming

As the main problems, low egg fertility, inadequate technical support, low market opportunity and disease were identified by 65.22, 23.91, 6.52 and 4.35% farmers, respectively (Figure 10).

Farmers' perception on prospects of turkey farming

According to 41.30, 28.26, 17.35 and 13.09% farmers' opinion main advantages of turkey rearing over other poultry species were low disease, high market price, low feeding cost and low mortality, respectively (Figure 10).

DISCUSSION

Turkey farming is a new farming enterprise in Bangladesh. Comparatively young population get involved with this farming and ownership of farming mostly belonged to male farmers. The present results on gender difference in the ownership of turkey agrees with the report of Yakubu et al. (2013) who observed a higher numbers of male than female among turkey keepers in Nassarawa state, Nigeria. Analysis of education data revealed that 100% farmer respondents received formal education ranges from less than Secondary School Certificate (SSC) to Master's Degree. The results indicate that participation of women in turkey farming is lower and turkey farmers are educated and most of them have prior experience. So, there is big possibility to flourish turkey farming by these farmers in near future.

The present study showed that although most of the farmers were rearing turkey for hatching egg and meat purposes, a large percent of farmers were raising turkey only for ornamental purpose. But there is a big opportunity to increase production for meat purposes because of its increasing demand to consumers of Bangladesh. Brant (1998) reported that different varieties of turkey are grown for pleasure and for competition at shows and exhibitions by hobbyists and fanciers in America. Most of the farmers were dependent on other farmers than government livestock offices for having technical support. The interviewed livestock officers informed that as turkey is a new species to them, so they do not have adequate awareness, knowledge and skill on it. It was observed that none of the farmers received any kind of training on turkey rearing. Average flock size of turkey was small because of newness of the enterprise. It was observed that some farmers were raising turkeys with other domestic fowl like chicken and duck in semi-intensive system. Most of the farmers reared both white and black turkeys. But white turkey is the most favored globally for meat (Osama et al., 2013). From discussion with farmers and farm observation it was assumed that the existing black and white birds would be the results of crossing between Broad Breasted Bronze, Broad Breasted White and Beltsville small white variety. They used sand, rice husk, wood shavings, coarse paper etc. as litter material. Even it was found that some farmers had not used any litter for mature turkey. It might be possible because of lower number of turkey in a flock. But in case of larger flock size adequate supply of suitable dry litter is a must to increase comfort and reduce disease incidence.

Most of the farmers used traditional broiler and layer brooding system for turkey. They used electrical brooder with bulbs and maintained temperature between 90°F to 95°F. But few farmers did not follow standard procedure for brooding which caused death of many poults at early stage. (Figure 11). Usually young poults by nature are reluctant to eat and drink in the first few days of life because of poor eye sight and nervousness; and for this reason force feeding is necessary during brooding period. But farmers were not aware about it. It was found that sometimes farmers fed poults manually without knowing the main reason. Results indicate that most of the farmers were not aware about turkey management in terms of housing, lighting, maintenance of hot and cold period.

Most of the farmers fed both homemade and commercial broiler and layer feed for feeding turkey. In case of homemade feed, they used a mixture of maize, wheat, broken rice and vegetables like cabbage, water spinach (*Ipomoea aquatic*), malabar spinachc (*Basella alba*) and grass. They allowed the turkey flock for foraging (Figure 12). Farmers were not aware about feed efficiency. But the importance of feed efficiency is high due to the high cost of feed, which represents approximately 70% of the total cost of a turkey production system (Wood and Willems, 2014). Most of the farmers supplied concentrate feed in the morning and evening. Supply of *ad libitum* water was not practiced in all the farms. It was observed that they did not follow nutrient requirement rules for turkey; even most of the farmers did not know it. But Turkey poults have high protein requirements for their first seven week (Robbins, 1983). Although farmers were rearing turkeys as breeder, they did not know about breeder ration requirement. In fact, their knowledge level on turkey feeding was very poor. Similar findings were found by Ojewola et al. (2002) in Nigeria and they reported that the farmers fed their breeder turkeys with different classes of commercial chicken feed probably because of insufficient knowledge of the levels of nutrient requirements of breeder turkeys.

Results indicate that prevalence of turkey disease was comparatively low. Most of the farmers had not used vaccines as preventive measure. Few farmers used vaccines mainly for New Castle disease, Fowl Pox and Fowl Cholera diseases. It seems that local turkeys are like indigenous chicken which are hardy and have high level of immunity against disease. Another reason of low disease prevalence might be that lower concentration of turkey farming in Bangladesh. Another reason of low use of vaccine might be that some farmers faced problems because of use of low potent vaccine. Being informed from victim farmers, other became either cautious or reluctant in using vaccine. But the poult stage was found to be the most vulnerable stage for disease attack of the local turkey. It was found that Fowl Pox mainly suffered turkey at poult stage (Figure 13). Peters (1997) reported that 74 (77.9%) out of the 95 interviewed turkey farmers had no record of disease attack in their flock in Nigeria. During in depth discussion on the issue it was found that most of the farmers did not follow deworming schedule for turkey as like chicken. Some experienced respiratory infection which might be due to Mycoplasma. Some farmers got weak poults with malformed legs which might be caused owing to improper temperature and humidity during incubation and poor nutrition of parents. Few farmers used ethno veterinary drugs such as the aloevera, turmeric etc. to treat sick turkey. There were divergent views on their efficacy in controlling and treating diseases and most of the knowledge on ethno-veterinary medicines was passed on orally to future generations.

The study revealed that price of adult turkey and poults were higher in Bangladesh in comparison to international market. The main reasons are that turkey subsector is still at the beginning stage in Bangladesh and in most cases turkeys were sold for ornamental purposes while some buyer bought also turkeys for farming as well as consumption purposes. Farmers bought egg for hatching purpose, so that they could raise turkey after incubating egg. Poults were sold without identifying their sex at the age from day old to 4-5 weeks of age. Selection and price of turkey depends on appearance, color, size and weight. Yakubu et al. (2013) showed that body size, egg number, hatchability, heat tolerance, body conformation and disease resistance were the traits of utmost importance for selection purpose among rural turkey farmers in Nasarawa state, Nigeria. However, there is absence of structured market for turkey in Bangladesh.

Weight of available adult tom, hen and egg in Bangladesh were comparatively lower than that of developed countries. This might be because of lighter varieties of turkey reared by the farmers of Bangladesh. The mating ratio found in the present study was higher than the ratio of 1: 2.75 reported by Yakubu et al. (2013) for turkey raised by local farmers at Nassarawa state in Nigeria. However, it was at the higher limit of the continuum (1.67-3.69) reported for native turkey breeders in the state of Me-hoecan, Mexico (Lopez Zavaha et al., 2008).

In fact, several factors like age, temperature, duration of light, mating problems, low nutrient etc., might be the reasons of low fertility of turkey hen in Bangladesh. But fertility is a very important measure of reproductive efficiency (Malecki et al., 2004). The problem of unfertilized eggs has long been identified as one of the most critical factors limiting the success of breeding programs and ranges from 10.0–98.2% (Dzoma and Motshegwa, 2009). Hatchability of eggs was lower because of lower fertility including insufficient knowledge of farmers on turkey breeding and egg incubating procedure. Although commercial livestock species completely dependent upon artificial insemination (AI) for fertile egg production (Juliet and Bakst, 2008), none of the respondent farmers used this technique.

Problems of turkey farming

Low fertility, hatchability and use of turkey reproduction technology

From the present study it was found that none of the farmers used AI technique and even they had not heard about it earlier regarding turkey breeding. In fact, adult body weight of tom has been increased over time due to advance researches and become too large to achieve natural fertilization. Anthony (2001) reported that modern White Turkey was developed for rapid growth rate through a selection process, which makes it so different from their wild ancestors that they are unable to mate naturally because of their heavy weight and AI has become necessary. Moreover, it has been reported that the hatchability of medium sized turkey eggs is better than that of small or large eggs (Kaygisiz et al., 1994). Age of the breeder is important factor which affects egg weight, internal and external quality egg, hatching performance and the quality of poult. It was reported that as hen age increases, the weight of egg increases and both shell quality and internal egg quality decrease (Erensayın, 2000). In addition to low egg yield, unsatisfactory egg fertility and hatchability constitute a major problem for turkey breeding enterprises (Ozcelik et al., 2009).

Inadequate access to technical information and support

The farmers did not have adequate access to necessary information regarding turkey rearing and in case of problems they did not get enough technical support from different government and non-government line agencies. This situation is also prevailed in other developing countries. Mbanasor and Saampson (2004) also reported that there was obvious lack of information on specific requirements for turkey production in Nigeria.

Low marketing facilities

Market of turkey is unlike broiler and layer in Bangladesh. There is absence of well-organized market for turkey and its products. No structured market value chain has been identified yet in Bangladesh. Farmers buy and sell turkey mainly through personal communication, Internet services (bikroy.com, Facebook etc.) and at the market of ornamental birds. Turkey selling problems is also identified in other developing countries as stated by Peters et al. (1997) in a study conducted on small holder local turkey production in Ogun State Nigeria, found that sale of turkeys were more during Christmas and festive period than other periods of the year. Although, turkey meat is being sold in department stores in capital city Dhaka, a large numbers of consumers were not habituated of taking turkey meat.

Poor housing

Farmers did not know the scientifically accepted space requirement for rearing turkey. They gave space on the basis of assumption. Moreover, they were not aware of about using of suitable litter materials and their management. Many farmers did not take special care during extreme hot and cold situation which ultimately hampered the production performance of birds.

Non availability of manufactured feeds and feeding standard

Feeds for turkey are not manufactured by any feed mill in Bangladesh. So farmers fed their turkeys by their homemade feed as well as a mixture of homemade and broiler/layer feed. They did not know the scientific requirement of energy, protein and other nutrients for different categories of turkey. Similar things was happened in Nigeria as reported that turkey production in Nigeria has largely remained at the smallholder level due to high cost of feed, inconsistency in feeding program, as well as lack of knowledge of the adequate levels of nutrient requirement (Ojewola et al., 2002). Although turkey is a good forager, some of the farmers did not know this fact so that they could not reduce feeding cost. Farmers did not have expertise to formulate balanced rations for turkey, thereby relying on rations originally formulated for layer and broiler chicken, with the assumption that chicken feed could bring same or better results. In this connection Etuk (2005) reported that lack of knowledge of limitations of feed ingredients used in turkey feeds leads to poor growth. But proper nutrition is a basic pre-requisite for successful poultry production (Kekeocha, 1984), to increase resistance to diseases and explore genetic potentiality.

Inadequate capacity building facilities

There is absence of opportunity for capacity building of turkey farmers in terms of receiving training, getting information, participating in workshop and seminar. As most of the concern stakeholders are not aware enough about turkey farming in Bangladesh, farmers are not getting required knowledge and skill. Therefore they are using traditional procedure for rearing turkey. But egg weight, fertility, hatchability and late embryonic mortality varied greatly between traditional and modern breeding management system (Lariviere et al., 2009).

Prospects of turkey farming

Adapted to the climate of Bangladesh

Turkey is a unique bird which is suitable for rearing in hot humid climatic condition like in Bangladesh. But due to unknown reasons it has not been explored in Bangladesh and other developing countries. In fact, turkeys are adaptable to wide range of climatic conditions and can be raised successfully almost anywhere in the world if they are well fed and protected against diseases and predators. The meat of turkey is considered by many as a luxury meat. Moreover, it has an aesthetic value due to their beauty (Ogundipe and Dafwang, 1980). For this reason turkey is becoming popular gradually in developing countries like in Bangladesh. Anandh et al. (2011) reported that commercial turkey farming is becoming popular in India.

Low disease prevalence

Turkey is more disease resistant in comparison to other poultry species like chicken, duck and quail. Mortality rate of turkey is very low in comparison to other poultry bird. Sampath (2012) reported that turkeys are resistant to Marek's and Infectious bronchitis and commonly encountered with other diseases like mycoplasmosis, fowl cholera, erysipelas and hemorrhagic enteritis. Farmers mostly do vaccination only for New Castle disease and Fowl cholera.

Low feeding cost

In fact, feed cost represents two thirds of the total costs in a poultry production system and consequently it would be valuable to identify animals that eat less but perform at the same level as their contemporaries. Turkeys are good foragers and it could reduce feeding cost. However, other poultry species such as geese and turkey can obtain added nutrients from forage because they are better able to digest fiber due to larger microbial population in their digestive tracts (Brad et al., 2010). On the other hand, Soliven (1984) reported that according to opinion of farmers of the Philippines, turkey rearing is profitable as long as the poults are properly fed and taken care of, and cost of production is cheap as almost 50% of the feed they eat is green vegetables and field grasses as supplement to commercial feeds.

Higher market demand

At present turkey market is limited to some particular customers as an ornamental bird as well as for meat purposes; and its price is higher than other poultry species. There are a good number of Christian people in Bangladesh who are fond of turkey meat in Christmas day. So there is huge opportunity to expand turkey market in Bangladesh as well as in abroad.

Alternative source of income and protein

While broiler meat market is facing problems of higher diseases and lower taste, turkey meat could be an alternative for consumers. So it could be an effective alternative source of protein. Moreover, this bird is quite suitable for uplifting livelihoods of small and marginal farmers as it can be easily reared in free range and under both intensive and semi-intensive system with little investment for housing, equipment and management. It may create good opportunity for unemployed youths to start farming and earn income. Turkey bird has a promising potential to be an alternative to livestock in meat production (Nixey, 1986). In the context of competitive feeding and management cost different countries searched such alternative source for protein. Okoruwa et al. (2006) reported that with the continued rise in the cost of production of cattle, sheep and goat, which are the primary sources of animal protein in Nigeria, it has become very necessary to explore efficient and less common but potential sources of animal protein for economic viability. Male and female British United Turkey reached, at 16 weeks of age, 14.60 kg and 10.25 kg, respectively (BUT, 2005). Moreover, the turkey has high dressing percentage that could amount to 87% of slaughter weight (Turkey management guide, 2012).

Opportunity to use artificial reproduction technique

As natural mating is not resulting fertile egg, so there is an opportunity to promote AI technique in turkey for the production of commercial hatching eggs. It will decrease cost for rearing more tom. It is reported that a well-developed pectoral muscle in turkeys, has prevented turkey toms to mate naturally (Etches, 1996), and making AI a necessity. Fertility could be improved in turkeys by using AI. In addition, efficiency of use of semen could be increased because each tom can produce enough sperm to inseminate approximately 30 hens (Childress, 2003).

Availability of educated farmers

Most of the surveyed farmers are comparatively educated and they were self-starter. So there is huge possibility to develop turkey entrepreneurs in Bangladesh. They will be able to receive technical knowhow on selection, brooding, breeding, feeding, housing etc. on turkey rearing easily.

CONCLUSION AND RECOMMENDATION

In fact, turkey production is still at primitive stage in Bangladesh which is characterized by poor housing, feeding, breeding and healthcare practices as well as inadequate availability of scientific information, technical services, credit facilities, training and marketing opportunities. So, to improve the turkey production, vigorous public extension service, training for farmers, opening of different avenues for research on turkey and identifying marketing strategies, are immediately needed in Bangladesh.

ACKNOWLEDGEMENTS

The authors gratefully acknowledge the support of the sub-project 'Enrichment of Research Capabilities for Postgraduate Studies on Advanced Animal Science' CP # 3314, W-2, HEQEP/UGC/World Bank to accomplish the study successfully.

REFERENCES

1. Adanza EG, 2006. Research Methods: Principles and Applications. Sta. Mesa, Quezon City: Rex Bookstore. pp: 81-82.
2. Anandh MA, PN Richard Jagatheesan, P Senthil Kumar, A Paramasivam, G Rajarajan, 2012. Effect of Rearing Systems on Reproductive Performance of Turkey. Veterinary World, 5: 226-229.
3. Anthony JS, 2001. Poultry, The Tropical Agriculturalist Series, Revised Edition, Center for Tropical Veterinary Medicine, University of Edinburgh, UK.
4. Begum IA, MJ Alam, J Buysse, A Frija and G Van Huylenbroeck, 2011. A comparative efficiency analysis of poultry farming systems in Bangladesh: A Data Envelopment Analysis approach. Applied Economics, 44: 3737-3747.
5. Besbes B, 2009. Genotype evaluation and breeding of poultry for performance under sub-optimal village conditions. World's Poultry Science Journal, 65: 260-271.
6. Brad B, T Elena and A Gernat, 2010. Maximizing Foraging Behaviour. University of Florida, IFAS Extension. pp: 12-13.
7. Brant AW, 1998. A brief history of the turkey. Worlds Poultry Science Journal, Vol. 54 (4): 365-373.
8. BUT, 2005. British United Turkeys .Commercial Performance Goals. 5th ed. British United Turkeys Ltd, Warren Hall, Broughton, UK.
9. Childress T, 2003. Talking Turkey: the care and feeding of your Thanks giving bird. Creative Loafing Media (37). Retrieved from http://edis.ifas.ufl.edu.
10. Dzoma BM, and K Motshegwa, 2009. A retrospective study of egg production, fertility and hatchability of farmed ostriches in Botswana. International Journal of Poultry Science, 8: 660-664.
11. Erensayın C, 2000. Scientific-Technic-Practical Poultry. Broiler breeding and hatchability. Vol.1. 2nd rev. ed. Nobel Publication, Ankara, Turkey (in Turkish, English abstract).
12. Etches RJ, 1996. Reproduction in Poultry. 1st Edn., Cambridge, CAB International. pp: 208- 233.
13. Etuk EB, 2007. Nutritional composition and feeding value of sorghum in turkey diets. Ph.D. Thesis, Federal University of Technology Owerri, Nigeria.
14. FAO, 2007. The State of the World's Animal Genetic Resources for Food and Agriculture, edited by B. Rischkowsky and D. Pilling. Rome.
15. FAOSTAT, 2012. Livestock Primary Production Data. Retrieved from http://faostat.fao.org.
16. Fiala N, 2008. Meeting the demand: An estimation of potential future greenhouse gas emissions from meat production. Ecological Economics, 67: 412-419.
17. Grimes J, J Beranger, M Bender and M Walters, 2007. How to raise heritage turkey on pasture. American livestock Breeds conservancy Pittsboro, NC27312 USA. Headquarters, 233 S.WAckes Drive, 11th floor Chicago, Illinois- 60606.
18. Juliet AL and MR Bakst, 2008. The current state of semen storage and AI technology. Biotechnology and Germplasm Laboratory, Beltsville Agricultural Research Center, Agricultural Research Service, USDA, Beltsville, MD, U.S.A.
19. Karki M, 2005. Growth, efficiency of utilization and economics of different rearing periods of Turkeys. Nepal Agricultural Research Journal, 6: 89-88.
20. Kaygisiz A, 1993. Research on hatchability characteristics of Pekin ducks. Turkish Journal of Veterinary and Animal Science, 17: 205-208.
21. Kekeocha CC, 1984. Pfizer poultry production handbook. 3rd edition...state Company/publisher and town.
22. Lariviere JM, C Michaux, F Famir, J Detilleux, V Verleyen, and P Leroy, 2009. Reproductive performance of the ardennaise chicken breed under traditional and modern breeding management system. International Journal of Poultry Science, 8: 446-451.
23. Lopez Zavala A, TC Monterrubio Rico, H Cano Camacho, O ChassinNoria, U Aguilera Reyes and MG Zavala Páram, 2008. Native turkey (Meleagris gallopavo) backyard production systems' characterization in the physiographic regions of the state of Michoacan, Mexico. TécPecuMéx. 46: 303-316.

24. Malecki IA, SWP Cloete, WD Gertenbach and GB Martin, 2004. Sperm storage and duration of fertility in female ostriches. South African Journal of Animal Science, 34: 158-165.
25. Mbanasor JA and A Sampson, 2004. Socio-economic Determinants of Turkey Production among Nigerian soldiers. International Journal of Poultry Science, 3: 497-502.
26. Nixey C, 1986. A comparison of growth and fat deposition of commercial avian species. 7th European Poultry Conference, Tours, Paris, 24-28.
27. Ogundipe SO and II Dafwang, 1980. Turkey Production in Nigeria. National Agricultural Extension Research and Liaison Service (NAERLS) Bulletin No. 22. pp: 2-22.
28. Ojewola GS, AD Udokainyang and V Obasi, 2002. Growth, carcass and economic response of local turkey poults to various levels of dietary energy. In: VA Aletor and GE Onibi (eds.).Increasing household protein consumption through livestock products. Proceedings of the 27th Annual Conf. of Nigeria Society for Animal Production, Akure, Nigeria, pp: 167-169.
29. Okoruwa VO, AE Obayelu and O Ikoyo-Eweto, 2006. Profitability of Semi-intensive Egg Production in South-West and South-South Zones of Nigeria. Nigerian Journal of Animal Production. 33: 118 -125.
30. Osama Elshiek Yassin, Salim Gibril, Al Hafiz Abdelrahman Hassan and Bushara A Bushara, 2013. A Study on Turkey (Meleagris Gallopavo) Raising in the Sudan, Journal of Applied and Industrial Sciences, 1: 11-15.
31. Owen OJ, AO Amakiri, EM Ngodigha and EC Chukwuigwe, 2008. The Biologic and Economic Effect of Introducing Poultry Waste in Rabbit Diets", International Journal of Poultry Science, 7: 1036-1038.
32. Ozcelik M, F Ekmen and O Elmaz, 2009. Effect of location of eggs in the incubator on hatchability of eggs from Bronze turkey breeders of different ages. South African Journal of Animal Science, 39: 214-222.
33. Peters SO, CON Ikeobi and OO Bamkole, 1997. Smallholder local turkey production in Ogun State, Nigeria, Proceedings of International Network for Family Poultry Development (INFPD), M-Bour, Senegal, December 9-13, 1997, pp: 197-208.
34. Robbins CT, 1983. Wildlife feeding and nutrition. Academic Press, Inc., New York. 343, pp: 33-37.
35. Sampath KT, 2012. Turkey farming: A profitable enterprise, National Institute of Animal Nutrition and Physiology, Adugodi Bangalore, India, 21: 2.
36. Soliven ME, 1984. Rural turkey rearing in the Philippines. Poultry International, 23: 94.
37. SPSS, 2013. Statistical Package for the Social Sciences, SPSS for Windows 10.00. Version 22.0 IBM Inc. Chicago, USA.
38. Turkey Management Guide, 2012. Central poultry development organization (SR), Hessarghatta, Bangalore 560088, Website: http://www.cpdosrbng.Kar.nic.in. Accessed in January 2016.
39. Wood BJ and OW Willems, 2014. Selection for improved efficiency in poultry, progress to date and challenges for the future. Proceedings of the 10th World Congress on Genetics Applied to Livestock Production, Communication No. 111.
40. Yakubu A, K Abimiku, IS Musa Azara, KO Idahor and OM Akinsola, 2013. Assessment of flock structure, preference in selection and traits of economic importance of domestic turkey (Meleagris gallopavo) genetic resources in Nasarawa state, Nigeria. Livestock Research for Rural Development, 25: 18.

18

MILK PRODUCTION PERFORMANCES OF CROSSBRED CATTLE AT THE VILLAGES OF JAMALPUR DISTRICT IN BANGLADESH

Maksuda Begum[1], Jahura Begum[1], Md. Kamrul Hasan Majumder[2], Mohammad Monzurul Hasan[3], Md. Shamsul Hossain[1] and Farukul Islam[1*]

[1]Bangladesh Agricultural University, Mymensingh-2202, Bangladesh; [2]Bangladesh Livestock Research Institute, Savar, Dhaka-1341, Bangladesh; [3]Milkvita, College Road, Islampur, Jamalpur, Bangladesh.

*Corresponding author: Farukul Islam, E-mail: farukkrishibid@gmail.com

ARTICLE INFO

Key words
Crossbred dairy cattle
Milk production
Villages
Jamalpur district
Bangladesh

ABSTRACT

Data on body measurements like BL (body length), CG (chest girth), WH (wither height) TM (test day milk production), PM (peak milk production), LP (lactation period), CFDC (cost for concentrate feed before test day milk production per cow), GGU (green grass used before test day milk production per cow) and husbandry practices, were collected from a total of 100 dairy cattle at the villages of Islampur upazila under the district of Jamalpur in Bangladesh from January to February 2017. Collected data were analyzed using SPSS software. The highest BL, CG, WH, CFDC and GGU were 175.71±0.42 cm, 161.74±0.24 cm, 123.82±0.11 cm 125.54±0.24 bdt/cow/day and 27.29±0.89 kg/cow/day, respectively. The highest amounts of TM, PM and LP were 9.36±0.60 liter/cow, 13.11±0.54 liter/cow and 247.14±1.47 days/cow, respectively. BL, CG and WH increased with increased of the age of crossbred cattle up to 150 months of old. Similarly, CFDC and GGU increased with the increased of the age of crossbred cattle up to 150 months of old. Farmers in the study area were not interested to keep breeding bulls for breeding purpose but they were using artificial insemination system to inseminate their cows and aware about production performance record of the inseminating bull. Farmers took health services from milk vita and sold milk to the same. They believed that dairy cattle rearing a profitable livelihood. TM, PM and LP increased with the increased of the age of crossbred cattle up to 150 months of old. However, increase of CFDC will increase TM, LP and PM but increase of GGU will increase only LP. On the other hand increase of CG will increase TM and increase of BL will increase LP and PM.

INTRODUCTION

Livestock is a vital component of agriculture and contributing about 3.10% to gross domestic products (GDP) and this is also contributing more than 6% of total foreign exchange earnings in Bangladesh (BER, 2015). Per capita meat and milk requirement are 120 gm/day and 250 ml/day, respectively but in Bangladesh, per capita availability of meat and milk are 102 gm/day and 120 ml/day, respectively (BER, 2013), while, here in Bangladesh annual production of milk is 3.46 million tons only but total annual requirement is 13.32 million tons (MT) of milk (BER, 2012). In Bangladesh, introduced exotic breeds like Holstein-Friesian, Jersy, Sahiwal, Hariana, Sindhi, Australian, Sahiwal-Friesian and improved varieties like: Pabna Cattle, Red Chittagong, Munshiganj Cattle, North Bengal Grey Cattle etc. are available. About 3.53 million milking cows, 2.61 million dry cows (cows without milk) out of total cattle population (23.4 million) in Bangladesh, while 2.13 million draught cattle, and 4.20 million improved cattle were also reported among these 23.4 million cattle (Banglapedia, 2014). Crossbred animals under Bangladesh condition contributes about 24% of the 6.9 million breedable cows and heifers (Huque et al., 2011) and Friesian x Local crossbred cows's milk production performance considerably improved over the decades (Bhuiyan, 2011). To improve genetic merit of the dairy animals, among the various mating systems crossbreeding of local non-descript cattle with exotic breeds of high genetic potential, is considered to be a rapid and effective method (Usman et al., 2012). Progeny tested bulls for dairy development in the country are in progress to add proven bull in dairy cattle industry (Bhuiyan et al., 2015). Sex, season of birth and genotype did not affect the traits like, birth weight, three-month, six-month, weaning weight and average daily gain of calves significantly (Rahman et al., 2015). Through supplementation of straw-based diets with vigna hay, feed intake, nutrient digestibility and live weight gain of indigenous cattle were improved, in Bangladesh (Hossain et al., 2015).

A number of researches have been conducted to evaluate reproductive and productive performances of crossbreds dairy cows under relatively controlled condition at research centers, government owned farms and in some urban and peri-urban dairy areas but, there are a few of such works conducted in rural areas especially under the small holder dairy farming areas. Moreover, research work addressing body measurements to relate milk production performances is scanty. So, the present study was designed and conducted to learn the correlation between milk production traits and body measurements of Friesian × local crossbred cattle under village condition in Bangladesh.

METHODOLOGY

Data on body measurements like BL, CG and BH; milk production performances like TM, PM and LP; feeds like CFDC and GGU and husbandry practices, were collected from a total of 100 dairy cattle at 28 small scale dairy cattle farms from eight villages of Islampur upazila under the district of Jamalpur in Bangladesh from January to February 2017. Data were collected using a pre-structured questionnaire by door to door visit from randomly selected farms in the study zone. Age range of cows was 28 to 150 months. All cows were divided into three age groups, where age group (CA) one belongs to age range from 28 months to 50 months, CA two belongs to age range from above 50 to 72 months while CA three includes cows of age range from above 72 to 150 months. Pedigree of crossbred cattle genotypes were like below (Table A).

The design of the study was unbalanced factorial in nature, because the observation numbers of different traits were unequal. The recorded data were stored on to the excel spread sheet and edited for further analyses. Then data were analyzed for having means through compare means menu, to obtain the relationship among the traits TM, PM, LP, BL, CG, WH, CFDC, GGU, CA and Calves ages, Pearson's correlation coefficient were used through correlate menu, and Duncan's Multiple Range Test (DMRT) were used for performing mean comparisons using the Statistical Package for the Social Sciences version 14.0 (SPSS, 2005).

Table A. Dam and sire genotype of enumerated cows were as follows

Dam genotypes	Frequency
Local	7
Local ×Friesian=Friesian crossbred 1	50
Friesian crossbred 1×Friesian crossbred 1=Frisian crossbred 2	39
Friesian crossbred 1×Friesian crossbred 2=Frisian crossbred 3	4
Total number of cows	100
Sire genotypes	**Frequency**
Local ×Friesian=Friesian crossbred 1	7
Friesian crossbred 1×Friesian crossbred 1=Frisian crossbred 2	87
Friesian crossbred 1×Friesian crossbred 2=Frisian crossbred 3	4
Friesian crossbred 2×Friesian crossbred 3=Frisian crossbred 4	2
Total number of cows	100

Body measurements were taken while animals were standing in a structure where it could not move around and keep calm. Body Length (BL) – was the distance from middle point between two horns in the centre of head to beneath the tail where anus open. Chest Girth (CG) – was measured as the minimal circumference around the body immediately behind the front shoulder, and Wither Height (WH) – was distance from the ground beneath the animal to the top of the withers directly above the centre of shoulder,

RESULTS

Body measurements

Body measurements varied significantly among the age groups of crossbred dairy cows (Table 1). BL significantly varied among the CA and the highest BL were documented for CA3 (175.71±0.42 cm) and these were followed by CA1 and CA2. But BL did not vary significantly between CA1 and CA2. CG significantly varied among the CA and the highest CG were documented for CA3 (161.74±0.24 cm) and these were followed by CA2 and CA1.

Table 1. Body measurements of enumerated cows at different ages

CA	BL	CG	WH
1	158.51[b]±0.33 (37)	146.91[c]±1.07 (37)	114.57[c]±0.11 (37)
2	158.06[b]±0.53 (35)	152.18[b]±0.64 (35)	115.97[b]± 0.60 (35)
3	175.71[a]±0.42 (28)	161.74[a]±0.24 (28)	123.82[a]±0.11 (28)
LS	***	**	**
Overall mean	163.17±0.04 (100)	152.90±0.54 (100)	117.65± 0.07 (100)

Note: BL-body length in cm, CG-Chest Girth in cm, WH-wither height in cm, CA 1- 28 months to 50 months, CA 2- above 50 to 72 months and CA 3-bove 72 to 150 months. LS=Level of significance, NS= Not Significant (P>0.05), *significance at 1.1 to 5%, **significance at 0.1 to 1%, ***significance at lower than 0.1% and [abc]Means with the different superscripts differed significantly within the column (P<0.05).

WH significantly varied among the CA and the highest WH were documented for CA3 (123.82±0.11 cm) and these were followed by CA2 and CA1.

Feeds and feeding of dairy cattle

The highest amount (125.54±0.24 bdt/cow/day) of cost for concentrate feed purchase before the test day milk production per cow (CFDC) was spent in age group 3 and these were followed by age group 2 and 1 (Table 2).

Table 2. Used concentrate feed and green grass for test day milk production.

CA	CFDC	GGU
1	97.30^c±0.83 (37)	14.76^c±1.73 (37)
2	99.14^b±0.43 (35)	15^b (35)
3	125.54^a±0.24 (28)	27.29^a±0.89 (28)
LS	*	***
Overall mean	105.85±4.85	18.35±0.95

Note: CFDC-concentrate feed cost in BDT for test day milk production, GGU-green grass (kg) used for test day milk production per cow, CA 1- 28 months to 50 months, CA 2- above 50 to 72 months and CA3-above 72 to 150 months.*significance at 1.1 to 5%, **significance at 0.1 to 1%,***significance at lower than 0.1% and ^{abc}Means with the different superscripts differed significantly within the column (P<0.05).

The highest amounts (27.29±0.89 kg/cow/day) of green grass were used before test day milk production per cow (GGU) in age group 3 and these were followed by age group 2 and 1.

Husbandry practices of crossbred dairy cattle

Male caves from all dairy cows were kept to sale for meat purpose and all cows were inseminated by AI workers. Most of the farmers (89.29%) collected information about milk production performances of inseminating bull's dam and sib before insemination (Table 3).

Table 3. Husbandry practices followed by the farmers

Traits	Opinion	
Male calf of cows use for	kept for sale for meat purpose	28 (100%)
	kept for breeding purpose	0
Insemination system	Use herd bull	0
	Artificial Insemination by AI worker	28 (100%)
Do you know milk production performance of	Yes	25 (89.29%)
inseminating bull's dam and sib before	No	3 (10.71%)
insemination of your cows?		
Do you use vaccine regularly for your cows	Yes	28 (100%)
	No	0
Do you use anthelmintics regularly for your	Yes	28 (100%)
cows	No	0
Receive health services from	Milk vita	27 (96.43%)
	Govt. Livestock Development Center	1 (3.57%)
Milk sale place	Milk vita collection center	25 (89.29%)
	Village and upazila market	3 (10.71%)
Is dairy cow husbandry a profitable livelihood?	Yes	28 (100%)
	No	0

However nearly all (96.43%) cattle farmers received health services for their cattle from Milk vita and the sold their cow's milk to Milk vita. All farmers used to vaccinate and administered anthelmintics regularly to their dairy cattle and they were in an opinion that dairy cattle husbandry is a profitable livelihood.

Peak milk production period
Most of the cows (87.00%) gave peak milk at her second month of a whole lactation period (Table 4).

Table 4. Peak milk production month

Traits		Number of lactating cows
Peak milk production in a whole lactation period	First month	9 (9.00%)
	Second month	87 (87.00%)
	Third month	4 (4.00%)

Milk production performances
The highest amounts of TM (9.36±0.60 liter/cow) were recorded for the cows with age group 3 and these were followed by age group 1 and 2 (Table 5). The highest amounts of PM (13.11±0.54 liter/cow) were recorded for the cows with age group 3 and these were followed by age group 1 and 2.

Table 5. Milk production performances at different ages

Cow age groups	TM	PM	LP
1	8.50±0.30 (18)	11.27[b]±1.49 (37)	221.22[b]±0.00 (37)
2	7.64±2.79 (22)	11.11[b]±0.76 (35)	212.89[b]±0.08 (35)
3	9.36±0.60 (25)	13.11[a]±0.54 (28)	247.14[a]±1.47 (28)
LS	NS	*	**
overall	8.54±1.39 (65)	11.73±1.11(100)	223.86±0.68(100)

Note: TM-test day milk production per cow in liter, LP-lactation period in days, PM-peak milk production per cow per day in liter, age group 1- 28 months to 50 months, age group 2- above 50 to 72 months and age group 3-bove 72 to 150 months.

The longest LP (247.14±1.47 days/cow) were recorded for the cows with age group 3 and these were followed by age group 1 and 2.

Correlation of TM with Calves age

Correlation between TM and calves ages (Table 6) was not significant and this was at minimal level (0.046).

Table 6. Correlation between calves age and TM

Parameter	Number	TM	Calves age (1 to 9 months)
TM	25	1	.046
Calves age (1 to 9 months)	25	.046	1

Note: Correlation is not significant at above 0.05 levels (2-tailed).

Correlation of milk production parameters with body measurements and feed consumption
TM was strongly and positively correlated (Table 7) with CG (medium level; 0.606) and CFDC (medium level; 0.620). LP was strongly and positively correlated with BL (lower level; 0.227), CFDC (lower level; 0.393), GGU (lower level; 0.232) and CA (lower level; 0.218).

Table 7. Correlations among milk production parameters with other traits

Parameters	N	CG	WH	CFDC	GGU	TM	LP	PM	CA
BL	100	0.457**	0.666**	0.078	0.237*	0.205	0.227*	0.201*	0.373**
CG	100		0.625**	0.046	0.305**	0.606**	0.153	0.167	0.343**
WH	100			0.020	0.301**	0.182	0.104	0.078	0.337**
CFDC	100				-0.016	0.620**	0.393**	0.616**	0.178
GGU	100					-0.057	0.232*	.073	0.420**
TM	65						0.234	0.715**	0.117
LP	100							0.516**	0.218*
PM	100								0.225*

Note: N-number of observations, BL-body length in cm CG-body circumference in cm, WH-body height in cm, CFDC-concentrate feed cost for test day milk production, GGU-green grass used for test day milk production, TM-test day milk production per cow in liter, LP-lactation period in days, PM-peak milk production per cow per day in liter, and CA-cow age in months. ** Correlation is significant at the 0.01 level (2-tailed), *Correlation is significant at the 0.05 level (2-tailed).

PM was strongly and positively correlated with BL (lower level; 0.201), CFDC (medium level; 0. 0.616) and CA (lower level; 0.225).

DISCUSSIONS

Body measurements of crossbred cows

BL significantly varied among the CA and the highest BL were found for CA3 and these were followed by CA1 and CA2. BL did not vary significantly between CA1 and CA2. But the distance from point of shoulders to the ischium of Brown Swiss (136.88±1.97 cm), Holstein (146.37±0.95 cm) and Crossbred (140.15±1.26 cm) were lower than present findings (Ozkaya and Bosket, 2008).CG significantly varied among the CA and the highest CG were documented for CA3and these were followed by CA2 and CA1 and the same were lower than Brown Swiss (180.25±3.38 cm), Holstein (189.36±1.73 cm) and Crossbred cattle (181.59±.66 cm) (Ozkaya and Bozkurt, 2008). WH significantly varied among the CA and the highest WH were observed for CA3 and these were followed by CA2 and CA1 and were in line with Ozkaya and Bozkurt (2008) who found, WH of Brown Swiss (123.45±1.40 cm), Holstein (132,60b± 0.66 cm) and Crossbred (127.95±1.14 cm). Body measurements like BL, CG and WH varied significantly among the age groups of crossbred dairy cows. Above discussions might be suggested that BL, CG and WH increased with increased of the age of crossbred cattle up to 150 months of old.

Feeds and feeding of dairy cattle

The highest amount of cost for concentrates feed purchase before test day milk production per cow (CFDC) was spent in age group 3 and these were followed by age group 2 and 1. Similarly, the highest amounts of green grass were used before test day milk production per cow (GGU) in age group 3 and these were followed by age group 2 and 1. On the other hand, Sanh et al. (2002) reported, body weight gain, milk yield and milk protein were increased as the ratio of concentrate feed was increased in the diet but milk fat was decreased. Above discussions might be indicative that CFDC and GGU increased with the increased of the age of crossbred cattle up to 150 months of old.

Husbandry practices of crossbred dairy cattle

Male caves from all dairy cows were kept to sale for meat purpose and all cows were inseminated by AI workers. Most of the farmers collected information about milk production performances of inseminating bull's dam and sib before insemination. However nearly all cattle farmers received health services for their cattle from Milk vita and they sold their cow's milk to Milk vita. On the contrary Islam et al. (2016) reported that farmers were not knowledgeable about modern health management and natural mating system was the main way of insemination at rural villages of Chapai Nawabganj district in Bangladesh. However, all farmers at present study area, used to vaccinate and administered anthelmintics regularly to their dairy cattle and they were in an opinion that dairy cattle rearing is a profitable livelihood. The above discussions might be suggested that farmers in the study area were not interested to keep breeding bulls for breeding purpose but they were using AI system to inseminate their cows and aware about production performance record of the inseminating bull. Farmers took health services from milk vita and sold milk to the milk vita. They believed that dairy cattle rearing a profitable livelihood.

Milk production performances

Most of the cows gave peak milk at her second month of a whole lactation period. The highest amounts of TM were recorded for the cows with age group 3 and these were followed by age group 1 and 2 and the same was lower than Hossain *et al.* (2016), who observed, daily milk yield in crossbred was 15.90±0.72 liter/cow. Similarly, the highest amounts of PM were recorded for the cows with age group 3 and these were followed by age group 1 and 2 but peak milk production was in line with Sarder *et al.* (2007) in Bangladesh (11.63±2.90 liter/day/cow). Moreover, the longest LP were recorded for the cows with age group 3 and these were followed by age group 1 and 2 and were in line with Islam *et al.* (2017) who found the longest Lactation Period in days (219.88±0.47) under body weight group 2. Above discussions might be indicative that TM, PM and LP increased with the increased of the age of crossbred cattle up to 150 months of old.

Correlation among the traits

Correlation between TM and calves ages was not significant. TM was strongly and positively correlated with CG and CFDC which might be indicative that the increase of CG and CFDC will affect the TM positively. LP was strongly and positively correlated with BL, CFDC, GGU and CA and this might be suggested that increase of BL, CFDC, GGU and CA will positively affect the LP. PM was strongly and positively correlated with BL, CFDC and CA and which might be pointed that increase of BL, CFDC and CA will positively affect the PM. However in a different study, Positive correlation of test day milk production with cost involved to feed the cow with concentrate feed on the day before test milk production (0.794) and green grass used the day before test milk production (0.453) were estimated (Islam *et al.*, 2017). So the above discussions might be indicative that, increase of CFDC will increase TM, LP and PM but increase of GGU will increase only LP. On the other hand increase of CG will increase TM and increase of BL will increase LP and PM. Similarly, increase of CA will increase LP and PM.

CONCLUSIONS

BL, CG and WH increased with increased of the age of crossbred cattle up to 150 months. Similarly, CFDC and GGU increased with the increased of the age of crossbred cattle up to 150 months. Farmers in the study area were not interested to keep breeding bulls for breeding purpose but they were using artificial insemination system to inseminate their cows and aware about production performance record of the inseminating bull. Farmers took health services from milk vita and sold milk to the same. They believed that dairy cattle rearing a profitable livelihood. TM, PM and LP increased with the increased of the age of crossbred cattle up to 150 months. However, increase of CFDC will increase TM, LP and PM but increase of GGU will increase only LP. On the other hand increase of CG will increase TM and increase of BL will increase LP and PM.

COMPETING INTEREST

There is no conflict of interest.

ACKNOWLEDGEMENT

We thankfully acknowledge the contribution of farmers of the villages under the upazila of Islampur of Jamalpur district in Bangladesh to support the research work during data collection.

REFERENCES

1. Banglapedia, 2014. Banglapedia, National Encyclopedia of Bangladesh. http://en.banglapedia.org/index.php?title=Cattle
2. BER, 2012. Bangladesh Economic Review (BER), Economic Division, Ministry of Finance. The Government of the People's Republic of Bangladesh, from http://www.mof.gov.bd
3. BER, 2013. Bangladesh Economic Review (BER), Economic Division, Ministry of Finance. The Government of the People's Republic of Bangladesh, from http://www.mof.gov.bd
4. BER, 2015. Bangladesh Economic Review (BER), Economic Division, Ministry of Finance. The Government of the People's Republic of Bangladesh, from http://www.mof.gov.bd
5. Bhuiyan AKFH, Rashid MM, Khan RA, Habib MA, Bhuiyan MSA, Faiz MA, 2015. Progeny tested bull production for dairy cattle development in Bangladesh. Bangladesh Journal of Animal Science, 44: 106-112.
6. Bhuiyan AKFH, Rashid MM, Khan RA, Habib MA, Bhuiyan MSA, Faiz MA, 2015. Progeny tested bull production for dairy cattle development in Bangladesh. Bangladesh Journal of Animal Science, 44: 106-112.
7. Bhuiyan AKFH, 2011. Progeny Tested Seed Bull Production: Progress towards National Dairy Development. Keynote paper presented at the 3rd National Seminar of Breed Up-gradation through Progeny Test Project held on 15 June 2011 at BIAM, Dhaka.
8. Hossain MS, Islam F, Rashid MHO, Leena SA, Sarker SC, 2016. Productive and Reproductive Performances of Holstein Friesian ×Local Crossbred and Pabna ×Pabna Cattle Genotypes. International Journal of Business, Social and Scientific Research, 4: 261-266.
9. Hossain MS, Miah MY, Khandaker ZH, Islam F, 2015. Effect of different levels of matikalai (*Vignamungo*) hay supplementation to straw-based diets on feed intake, digestibility and growth rate of indigenous cattle. Livestock Research for Rural Development, 27 (2).
10. Huque KS, Dev GK, Jalil MA, 2011. High yielding dairy breed development in Bangladesh- limitations and opportunities. Paper presented in international workshop on "High yielding dairy breed development in Bangladesh" held at Bangladesh Livestock Research Institute, Dhaka from 28-29 September.
11. Islam F, Alam MP, Hossain MS, Leena SA, Islam MR, Hasan SMR 2016: Prospects and challenges of homestead cattle production in the villages of ChapaiNawabganj district in Bangladesh. International Journal of Agronomy and Agricultural Research, 9: 44-50.
12. Islam F, Faruque MO, Ferdous F, Joya SH, Islam R and Hossain MS, 2017. Effect of genetic and non-genetic factors on milk production performance of Holstein-Friesian×Local crossbreds at the villages of Noakhali district in Bangladesh. Research in Agriculture, Livestock and Fisheries, 4: 21-28.
13. Ozkaya S and Bozkurt Y 2008. The relationship of parameters of body measures and body weight by using digital image analysis in pre-slaughter cattle. Archiv Tierzucht, Dummerstorf, 51: 120-128.
14. Rahman SMA, MSA Bhuiyan and AKFH Bhuiyan, 2015. Effects of genetic and non-genetic factors on growth traits of high yielding dairy seed calves and genetic parameter estimates. Journal of Advanced Veterinary and Animal Research, 2: 450-457.
15. Sanh MV, Wiktorsson H, Ly LV, 2002. Effects of Natural Grass Forage to Concentrate Ratios and Feeding Principles on Milk Production and Performance of Crossbred Lactating Cows. Asian-Australasian Journal of Animal Science, 15: 650-657.
16. Sarder MJU, MM Rahman, S Ahmed, MR Sultana, MM Alam, MM Rashid, 2007. Consequence of dam genotypes on productive and reproductive performance of dairy cows under the rural condition in Bangladesh. Pakistan Journal of Biological Sciences, 10: 3341-3349.
17. SPSS, 2005. Windows for version-14.0. Release on 27.10.2005. (Microsoft Corp. 1998). Trends SPSS Inc., Michigan Avenue, Chicago, IL. 19-182.
18. Usman T, G Guo, SM Suhail, S Ahmed, L Qiaoxiang, MS Qureshi and Y Wang, 2012. Performance traits study of Holstein Friesian cattle under subtropical conditions. The Journal of Animal and Plant Sciences, 22: 92-95.

19

IDENTIFICATION OF TYPES OF BUFFALOES AVAILABLE IN KANIHARI BUFFALO POCKET OF MYMENSINGH DISTRICT

Md. Mahmodul Hasan Sohel[1]* **and Md. Ruhul Amin**[2]

[1]Department of Animal Science, Faculty of Agriculture, Erciyes University, Kayseri-38039, Turkey; [2]Department of Animal Breeding and Genetics, Faculty of Animal Husbandry, Bangladesh Agricultural University, Mymensingh-2202, Bangladesh

*Corresponding author: Md. Mahmodul Hasan Sohel, E-mail: sohel.mmh@gmail.com

ARTICLE INFO

Key words

Buffalo Morphometric
Characterization
Kanihari
Buffalo-pocket

ABSTRACT

The buffaloes are reared by many races under diverse agro-climatic conditions of Bangladesh. The buffaloes of Bangladesh are mostly indigenous in origin. Both the swamp and river type buffaloes are found in Bangladesh and they can be found throughout the country. However, their concentration is higher in coastal part, Meghna-Ganga and Jamuna-Brahamaputra flood plain, subsequently forming buffalo pockets. This study was conducted to identify the types of buffaloes and the sources of breeding buffaloes in one of those buffalo pockets called Kanihari buffalo pocket situated in Mymenshing district. Direct interviewing method was used to collect the data from the owner of the buffaloes. According to the body shape, coat color and horn pattern, buffaloes of this region were categorized into different categories. Abundant natural green grasses in the river bank of old Brahammaputra gave this area as a shape of a buffalo pocket. This pocket does not have any distinct breed and the buffalo population is mostly mixed and exotic. Introduction of swamp germplasm occurred when buffalo cows are temporarily migrated to Bathan area of Sylhet. Morphometric characteristics (coat color, horn pattern and body appearance) of Type-1, Type-2 and Type-3 buffaloes are similar to Murrah group (Murrah and Nili-Ravi), Surti group and indigenous river type buffaloes, respectively. This is the first study which identifies and morphologically characterizes the buffalo population in Kanihari buffalo pocket; however, in depth genotypic study is required in order to identify the origin or breeds available in this area.

INTRUDUCTION

The buffalo (*Bubalus bubalis*) population in the world is estimated at 185.28 million head which is spread over 42 countries, of which 179.75 million buffaloes (97%) are found only in Asia (FAO, 2005). India has the largest buffalo population of 105.1 million and they comprise approximately 56.7% of the total world buffalo population. Bangladesh has around 830764 buffalo head owned by 270228 holdings representing 1.52% of the total holdings in the country (Faruque, 2003). During the last 10 years, the world buffalo population increased by approximately 1.49% annually. Worldwide interest has been developed on this species not only as a purveyor of animal protein in terms of meat and milk for human consumption, but also their adaptability to harsh environment. Furthermore, strong image linking between buffaloes with nature and environment-ecological equilibrium, particularly in some of the world most marginal land shows this species as one of the most viable alternative of intensive cattle husbandry system (Hayashi et al., 2005).

Cattle and buffaloes are two important species supplying more than 99% of total milk consumed in Bangladesh. Bangladesh is a densely populated country where in each km^2 approximately 964 people are living (BBS, 2011). To fulfill the growing demand of milk for a large population several attempts were made to mitigate the deficiency as well as to increase the total milk production of the country. However, most of the attempts were made to improve the genetic potential and production efficiency of cattle. These include small holding dairying, good management practice and introduction of crossbreeds to the domestic population. In addition with high yielding production potential crossbreed cows have several disadvantages including high quality feed requirement, susceptible to disease and low fertility. Therefore, it became necessary to choose alternatives for sustainable improvement of the consistent production. In this regard, buffalo can serve better than cattle. Moreover, buffaloes have a number of advantages over cattle. For instance, growing buffaloes may utilize coarse feed more efficiently than cattle (Singh and Mudgal, 1971), have more disease resistance ability and produce more solids in milk (Dubey et al., 1997) and require less management inputs. It has also been reported that buffalo holds strategic place in overall livestock economy of Bangladesh and serves three important purposes such as milk, meat and drought power supply (Ghaffar et al., 1991).

Dairy buffaloes are observed sparsely throughout the country. It has been reported that they are found to be concentrated in Meghna-Ganga flood plain as well as Brahammaputra-Jamuna flood plain (Faruque et al., 2007b). In addition, there are few buffalo pockets in Bangladesh including Coastal area, Sylhet haor area, Sugar cane belt of Jamalpur and Kanihari buffalo pocket in Trishal upazila of Mymenshing district. Among all the buffalo pockets particularly Kanihari buffalo pocket draw our attention because it has been reported that the people of Kanihari area used to keep buffaloes beside cattle mainly for milk production over hundreds of years and the genetic and phenotypic evaluation of these buffalo population has not been done yet. Higher concentration of dairy buffaloes in this area may be due to higher abundance of green grasses, milk marketing opportunity and higher price of buffalo milk than cow milk. However, this pocket does not have any particular breed and the owners do not follow any specific breeding plan. During the breeding season breeder buffaloes with different genetic makeup are reported to come from India through various rout. Moreover, introduction of swamp germplasm to the population has higher chance during the time when buffaloes are temporarily migrated to Sylhet bathan area (Faruque 2007a).

Therefore, aimless breeding of these animals were taking place causing no visible genetic improvement rather a gradual decrease in production can be noticed. Neither government nor NGOs were found to be involved in this area to improve the current condition of these animals in the pocket. Farmers of this area are often confused to choose the correct breed or type of buffaloes for breeding purpose which can serve them properly. Under these circumstances, a thorough study on the resource and potential of the pocket is appeared to be very essential in order to have a clear scenario of the genetic resources and production potential of the milk buffaloes. Considering these facts, current study was conducted i) to identify the routes of breeder buffaloes/seeds in the pocket and ii) to determine the breeds or types of buffaloes available in this area by morphometric characterization.

MATERIAL AND METHODS

The current experiment was conducted to identify the types of buffaloes on the basis of their phenotypic appearance. Data was collected within a period of 12 months starting from May 2007 to April 2008. In order to collect data a door to door survey was carried out. To know the origin of a particular buffalo, information was collected from its owner using a standard questionnaire.

Location of the study area

Kanihari union at Trishal upazila was selected for the present study. The influencing factors in choosing Kanihari union as study area were- localized concentration of buffaloes gave a form of buffalo pocket in this area, no substantial study was conducted previously in this area and it was well communicated which might be helpful for the researcher in collecting the data required for the research. First the study area was segmented into three parts namely Area 1, Area 2 and Area 3 in order to ease collection of data. Area 1 consists of village Jilki, Gobindapur, Balidia and Betra whereas Area 2 consists of village Kuista, Baghadaria and Garpara. On the other hand Area 3 consists of village Thapanhala, Dewpara and Sultanpur.

Number of animals

Animals were randomly selected from the whole study area. Data from 120 animals, of which 45 from Area 1, 40 from Area 2 and 35 from Area 3 were collected using a suitable questionnaire.

The survey schedule

The questionnaire was prepared in accordance with the objectives of our study. It was designed in a simple manner to obtain accurate and as much as information possible from the farmers. The questionnaire comprised of following major items-
   a) General information about the selected livestock owner i.e. occupation, income, income   from livestock and family size etc.
   b) Purchase and breeding history of the buffalo
   c) Live weight
   d) Phenotypic and morphologic assessments i.e. coat color, horn pattern etc.

Data collection, record keeping and statistical analysis

Door to door interviewing method was used to collect the data from specific buffalo owners. Information given by the owner was recorded in a record book and kept for further analysis. After collection, data were processed, tabulated and analyzed using Statistical Program for Social Science (SPSS, Version 16.0) computer program.

RESULTS AND DISCUSSION

A glimpse on Kanihari Buffalo pocket

Kanihari Union (terminal local govt. unit) is located 25 km south from Mymensingh city and 100 km north from capital Dhaka along west side of old Brahamaputra river. The extended sandy and loamy basin of old Brahamaputra River provided this area as an excellent grazing land particularly in wet season. The union consists of 12 villages, among them Jhilki, Sultanpur, Kuista and Chor-baghadaria possesses highest number of buffaloes. There are around 285 household rearing approximately 850 buffalo heads. It is reported that from hundred years back people of this area used to keep buffaloes beside cattle mainly for milk production. Higher concentration of dairy buffaloes in this area might be due to firstly, abundance of natural grasses which grow over the sandy and loamy basin of old Brahammaputra river and buffalo can utilize these grasses better than cattle. Secondly, high demand and milk marketing opportunity in Mymensingh Sadar mainly for raw consumption and production of milk products. Milk traders used both rail and water route to bring the milk from Kanihari to Mymensingh city. Thirdly, because of buffalo milk contain higher total solid and fat, typically buffalo milk sells at almost double price than cow milk. It is a general observation that buffaloes in Kanihari produced

3-5 times higher amount of milk each day than cattle. While daily milk production of cattle is 1-2 lit, buffalo produce 5-8 lit milk/day. Furthermore, the price of buffalo milk is higher than cow milk because it contain higher amount of total solids and fats. It is important to note that farmers who rear buffaloes side by side of their main occupation, they are generating 35-40% more of their total income. Because of these facts buffaloes are popular in Kanihari and it became an integral component of livestock agriculture in this area.

Sources of buffalo stock and breeding policy

Replacement and introduction of exotic buffaloes can occur in several ways. We observed that there are at least four main sources from where buffaloes are get into this area-

a)   Locally born replacement calves
b)   Indian buffaloes coming through traders via different routes
c)   Buffaloes from Sylhet Bathan area
d)   Buffaloes from Myanmar (Barma), reported by some farmers

Farmers used to buy buffaloes in different recognized market such as Gabtoli Market, Kalimela in Dinajpur district. Some farmers buy and sell their buffaloes when they are migrated temporarily in the Bathan are of Sylhet. Sometime traders bring their buffaloes directly in Kanihari and sell to farmers.

We found that Kanihari buffalo pocket does not have any distinct breed and the buffalo population is mostly mixed. It is alarming to note that the owners are not following any specific breeding plan. As a result aimless and haphazard breeding of buffaloes were taking place in this area. They are typically inseminated by locally available bulls or by swamp buffaloes when they are temporarily migrated to bathan area causing introduction of swamp germplasm in this area. Therefore, there is a gradual decrease in milk production performance was observed in locally born replacement crossbreeds (River × Swamp) or (Exotic × Deshi).

An overview of Types of buffaloes observed in Kanihari Buffalo Pocket

We collected data from 120 owners which are corresponding 120 adult buffaloes. Among the buffaloes investigated there are at least four types of buffaloes were found to be identified in this area and subsequently, they were grouped as Type-1, Type-2 and Type-3. Each type of buffaloes has distinct coat color, special markings, specific horn pattern and different production potential. An overview of different types buffaloes are listed in Table 1. Phenotypic characterization generally refers to the process of identifying distinct breed populations and describing their external and production characteristics within a given production environment. The term "production environment" is here not only refer the natural environment but also include the management practices and as well as social and economic factors. It is well known that each breed has distinct characteristics and it is possible to assess the type or breed of buffalo by analyzing its phenotypic characteristics. Keeping these facts in mind, in order to have a clear idea about the buffalo population in Kanihari Buffalo Pocket different phenotypic aspect were analyzed including body appearance, coat color and horn pattern.

Table 1. Morphological characterization buffaloes in Kanihari buffalo pocket

Group	Local name	Coat color	Horn pattern	% of total population	Farmer's preference
Type-1	Gurjuti	Jet black, soft, smooth & scanty hair	Short, tightly curled & forming coil	20.14	35
Type-2	Nepali	Fawn or black, rusty brown hair and black or white switch	Short to medium in size and sickle shaped	43.24	45
Type-3	Deshi (milk type)	Black	Comparatively large, no definite shape	36.62	20

Body appearance: Type-1 buffaloes have sound build, massive and heavy barreled shaped body with comparatively short head but long face and neck and udder is well developed. They have long tail which in generally reaching to the fetlock joint. Similar type of body structure and appearance was observed in Murrah breeds (Sangwan, 2012). On the other hand Type-2 buffaloes are medium in size, well-shaped body with a wide and long head and the udder is well developed and extends well forward up to the naval flaps. These body characteristics are typical for Surti breeds (Sangwan, 2012). Type-3 buffaloes are comparatively smaller in size than Type-1 and Type-2 buffaloes. And they have wide forehead with a long face. The udder is smaller in size with longer teat compare to Type-1 and -2.

Coat color: Each breed has distinct coat color and hair pattern. It has been observed that the coat color of the buffaloes in this area was found to be ranged from fawn/light black to jet black. The coat color of Type-1 buffalo is predominantly jet black with soft, smooth and scanty hair. Coat color of Type-2 buffaloes was ranged from black, rusty brown to reddish. Among the Type-2 buffaloes, we found around 60% animals have rusty brown coat color and 40% animals have black coat color in this area. On the other hand, Type-3 buffaloes have black coat color. Interestingly, one albino buffalo calf was found in Kanihari buffalo pocket, however, its genetic background was unknown. It has been reported that, wide range of skin and coat color can be observed among the various breeds and types of water buffaloes throughout the world. Coat color varies with the age, husbandry practice and in relation to the degree of exposure to the solar radiation. The coat color of most breed darkens gradually with the advancing of age, especially after the first five years of life (Macgregor, 1941). Pundir and his colleagues mentioned that Surti buffaloes are usually copper colored with scanty hair which is black at the roots and reddish brown at the tips and some time the hair is completely rusty brown (Pundir et al., 2000). It has been reported by Faruque (1989) that the coat color of buffaloes in Bangladesh is jet black in western region, deep grey to light grey in central and eastern region. The coat color of buffaloes in south east region is predominantly grey (Faruque and Amin, 1994).

Special markings: Chevron marking on the chest was found in 17.39% of the buffaloes in Kanihai. In general, typical Surti buffalo has two strips or chevrons on its chest. White spot in forehead was found in 12.55% of buffaloes, while white tail switch was observed in 62.15% of the buffaloes. White spots on forehead and tail-switch are characteristics markings for Surti breed. It has been reported by Faruque (2003), a typical Bangladeshi indigenous buffalo is river type and sometime white tail-switch can be observed. However, white markings on forehead and tail switch can be seen also in Murrah breeds but it is not preferred. Nili-Ravi buffaloes are similar to Murrah in almost all characteristics except the white markings on extremities. They have usually wall eyed and have white markings on forehead, face and tip of the tail (Banerjee, 2011).

Horn pattern: Buffaloes in Kanihari have a wide range horn pattern starting from short curled to long shaped horns. For example, Type-1 buffaloes possess short, tightly and spirally curved horns. This type of horn pattern has been reported for the characteristic phenotype of Murrah buffalo breed (Rife, 1962). Type-2 buffaloes have flat horns with medium to short in length, sickle shaped and are directed downward and backward, and then turn upward. Although, Type-3 buffaloes have no definite horn shape; however, they possess comparatively large horns which are generally grow backward then upward and then forward forming hook-shaped structure. Many river buffaloes have spirally curled horns. They are tightly coiled in Murrah, Nili-Ravi and Kundi, downward curving in the Jafrabadi (Freed et al., 1981). The horns of Murrah buffaloes are different from other breeds, in general they are short, tight, turning backward and then upward and finally curving spirally inward. The horns get loosened slightly but spiral curves increased with the advances in age (Cruz-Cruz et al., 2014). The horns of Surti buffaloes are of medium length and sickle-shaped with flat and transverse corrugations. They are directed downward and backward and then turn upward at the tip to form a hook (Banerjee, 2011). Crosses between Murrah and Swamp produce medium sized slightly curled horns (Mingala et al., 2007). The horns of the buffaloes in the Southeast coastal areas of Bangladesh are of crescentic shape. This pattern of the horns is also found in most buffaloes in the south and some buffaloes in the south west. Small, spiral curled pattern horns are found in some cases in the south and south west coastal region (Faruque, 1994). In Brahammaputra-Jamuna flood fed area the horns of dairy buffaloes resemble to those of river type; curly horns through in few cases hanging horns were also observed (Fahimuddin, 1975; Mahadeven 1992). In Meghna Ganges flood plain areas most of the dairy buffaloes had crescentic horn which is the characteristics of swamp buffalo (Fahimuddin, 1975). The horn pattern of buffaloes in Tangail district ranged from spiral to almost crescentic (Hussen, 1990).

Cockrill described the buffaloes of Bangladesh as non-descriptive types. No substantial attempts were made to identify and characterized native buffalo stocks before 1980. After that attempts were made by Faruque (1990); Faruque and Amin (1995) to characterize the native and exotic buffalo population in different region of Bangladesh. Faruque (1994) reported that the coat color of buffaloes in Mymensing region varied from jet black to light grey. They also mentioned some buffaloes are with white stockings, tail switch and spot in forehead without mentioning their frequency. The results of this study provide a clear picture of different types of buffaloes in Kanihari buffalo pocket by analyzing their morphologic characteristics.

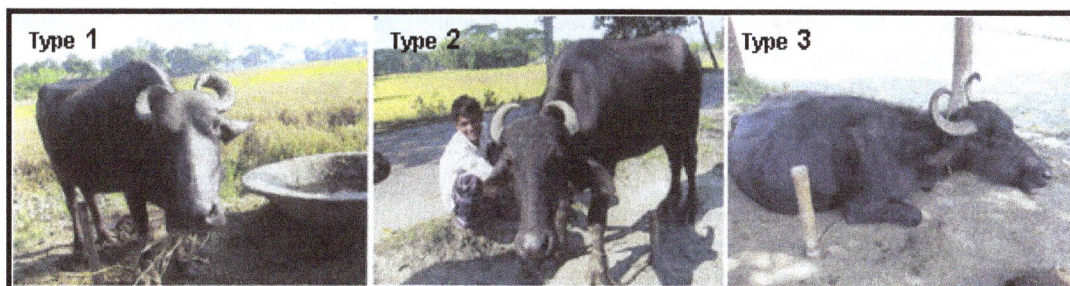

Figure 1. Horn pattern of three types of buffaloes available in Kanihari buffalo pocket

CONCLUSION

Our study suggests that the buffaloes in this region are an integrated part of livestock agriculture which has a strong impact on socio-economic condition of owners. Moreover, it can be concluded from the observed results and discussion that the morphometric characteristics of Type-1, Type-2 and Type-3 buffaloes are similar to Murrah group (Murrah and Nili-Ravi), Surti group and indigenous river type buffaloes, respectively. We also found few buffaloes with the characteristics of indigenous swamp buffalo in this region, however their frequency is very low (total of 5 in the study area) and they are mainly used for drought purpose. In addition to this study, we strongly suggest in depth studies in genetic level in order to identify the genetic origin of these buffaloes in study region.

ACKNOWLEDGEMENT

The authors would like to thank Prof. Dr. Shamsuddin and his team for their help during this study. The authors would also like to thank Masreka Khan for her contribution during preparation and editing the manuscript.

REFERENCE

1. Banerjee GC, 2011. A Textbook of Animal Husbandary, 8th ed., p. 703. Published by Oxford & IBH Publishing Co. Pvt. Ltd., New Delhi, India.
2. BBS, 2011. 5th National Census. http://www.bbs.gov.bd/home.aspx.
3. Dubey PC, CL Suman, MK Sanyal, HS Pandey, MM Saxena and PL Yadav, 1997. Factors affecting composition of buffalo milk. Indian Journal of Animal Science, 67: 802.
4. Fahimuddin M, 1975. Domestic water buffalo, Oxford & IBH Publishing, New Delhi, India.
5. FAO, 2005. Buffalo production and research. ftp://ftp.fao.org/docrep/fao/010/ah847e/ah847e.pdf.
6. Faruque MO and MI Hossain, 2007b. The Effect of Feed Supplement on the Yield and Composition of Buffalo Milk. Italian Journal of Animal Science, 6: 488-490.

7. Faruque MO and MR Amin, 1995. Indigenous buffaloes in the coastal area of Bangladesh. Part II. Productivity of indigenous buffaloes in the south western coastal area. Bangladesh Journal of Training and Development, 8: 138-140.

8. Faruque MO, 1994. The indigenous buffalo in the Mymensingh district of Bangladesh. Buffalo Journal, 2: 91-100.

9. Faruque MO, 2003. Buffalo production system in Bangladesh. Proc. of the Fourth Asian Buffalo Congress, New Delhi, India, 25 to 28 Feb.: 31-35.

10. Faruque MO, 2007a. The Genetic diversity of Bangladeshi Buffaloes. Italian Journal of Animal Science, 6: 349-352.

11. Faruquel MO, MA Hasnath and NU Siddique, 1990. Present status of buffaloes and their productivity in Bangladesh. Asian Australian Journal of Animal Science, 3 (4): 287-292.

12. Freed SA and RS Freed, 1981. Sacred Cows and Water Buffalo in India: The Uses of Ethnography. Current Anthropology, 22: 483-502.

13. Ghaffar A, MI khan, MA Mirza and WH Prizada, 1991. Effect of year and calving season on some traits of economic importance in Nili-Ravi buffaloes. Pakistan Journal of Agricultural Research, 12: 217.

14. Hayashi Y, KL Maharjan and H Kumagai, 2005. Feeding Traits, Nutritional Status and Milk Production of Dairy Cattle and Buffalo in Small-scale Farms in Terai, Nepal. Asian-Australasian Journal of Animal Science, 19: 189–197.

15. LA de la Cruz-Cruz, I Guerrero-Legarreta, R Ramirez-Necoechea, P Roldan-Santiago, P Mora-Medina, R Hernandez-Gonzalez and D Mota-Rojas, 2014. The behaviour and productivity of water buffalo in different breeding systems: a review. Veterinary Medicine, 59: 181-193.

16. Mahadevan P, 1992. Distribution, ecology and adaptation of river buffaloes. In: Buffaloproduction, production-system approach. (Eds. M. H. Tulooh, J.H.G. Holmes World Animal Science, c6), Elsevier Scientific Publications, Amsterdam, Netherlands. p. 1–58.

17. Mcgregor R, 1941. The domestic buffalo. Veterinary Research, 53: 443-450

18. Mingala CN, R Odbileg, S Konnai, K Ohashi and M Onuma, 2007. Molecular cloning, sequencing and phylogenetic analysis of inflammatory cytokines of swamp type buffalo contrasting with other bubaline breeds. Comparative Immunology, Microbiology and Infectious Diseases, 30: 119–31.

19. Prasad RMV, K Sudhakar, ER Rao, BR Gupta and M Mahender, 2010. Studies on the udder and teat morphology and their relationship with milk yield in Murrah buffaloes. Livestock Research for Rural Development, 22: 20.

20. Pundir RK, G Sahana1, NK Navani, PK Jain1, DV Singh, Satish Kumar and AS Dave, 2000. Characterization of Mehsana Buffaloes in India. Animal Genetic Resources Information, 28: 53-62.

21. Rife DC, 1962. Color and horn variations in water buffalo. The inheritance of coat color, eye color and shape of horns. Journal of Heredity, 53: 239-46.

22. Sangwan ML, 2012. Analysis of Genetic Diversity of Indian Buffalo Breeds by DNA Markers. Journal of Buffalo Science, 1: 91-101.

23. Singh, BB and BP Singh, 1971. Comparative study of life-time economics of Haryana cows versus Murrah buffaloes. Indian Veterinary Journal, 48: 485-489.

20

IMMUNE RESPONSE OF FOWL CHOLERA VACCINE PRODUCED AT BANGLADESH AGRICULTURAL UNIVERSITY IN FARM LEVEL

Md. Abdus Sattar Bag*, Md. Mansurul Amin, Md.Bahanur Rahman, Yasir Ahammad Arafat, Md. Salim, Imrul Hasan Rasel and Ummay Habiba Majumder

Department of Microbiology and Hygiene, Faculty of Veterinary Science, Bangladesh Agricultural University, Mymensingh-2202, Bangladesh

*Corresponding author: MA Sattar Bag; E-mail: bag201@yahoo.com

ARTICLE INFO

Key words

Fowl cholera
Vaccine
Immunogenicity
PHA titres
Antibody

ABSTRACT

The work was carried out to determine the immunogenicity of fowl cholera vaccine (FCV) produced by Livestock and Poultry Vaccine Research and Production Centre (LPVRPC) at Bangladesh Agricultural University (BAU). A total five hundred of seven-week old Hy-sex chickens (both white and brown) were vaccinated @ 0.5 ml of 2.93×10^8 CFU through subcutaneous route in each selected groups such as A1, A2 and A3; and B1, B2 and B3. Booster dose was provided at 13 weeks of age in group A3 and B3. Group C was kept as unvaccinated/control. Postvaccination sera were collected at different time schedule from all the groups of birds and antibody against fowl cholera were determined by Passive haemagglutination (PHA) test. At 4 weeks of primary vaccination (11 weeks aged birds) the mean PHA titres of sera were 96.00±34.21 and 96.00±34.21in group A1 and B1 respectively. On the other hand mean PHA titres at 5-weeks following vaccination (12 weeks aged birds) were 88.00±33.12 and 96.00±34.21in group A2 and B2, respectively. After 4 weeks of booster vaccination the mean PHA titres were 104.00±33.12in A3 and 104.00±33.12 in B3 group. The mean PHA titres in chickens of unvaccinated control group C was <4±0.00.Fowl cholera vaccine prepared at LPVRPC induced a good level of immunity at the farm level and it was also demonstrated that booster (secondary) vaccination is essential to develop protective level of immunity.

INTRUDUCTION

Poultry industry is an excellent agribusiness with its tremendous development during the last decades (1996-2006) in Bangladesh (Rahman, 2003). However, a number of infectious diseases of different etiologies such as bacteria, virus, fungi, mycoplasma etc. are found to be the most leading causes of economic loss often discouraging poultry rearing in this country (Das *et al.,* 2005). Among the bacterial diseases, fowl cholera (FC) is a major threat to the poultry industry. It is a contagious acute fatal septicemic disease of various domestic and wild bird species (OIE, 2004).Vaccination is practiced as preventive measures in Bangladesh like other countries of the world to reduce the incidence of the disease. Michael *et al.,* (1979) suggested that a local strain of higher immunogenic value should be selected as vaccine strain for preparation of a prophylacticbacterin. Fowl cholera vaccines (FCV) are made available in Bangladesh by a number of Pharmaceutical companies, Livestock Research Institute (LRI) of Department of Livestock Services (DLS) and Livestock and Poultry Vaccine Research and Production Centre (LPVRPC) (erstwhile known as poultry biologics unit) of Department of Microbiology and Hygiene, Bangladesh Agricultural University (BAU). The volume of FCV, an adjuvantedkilledbacterin produced at LPVRPC is on increasing demand. For this, it was thought plausible to investigate the on farm immunogenicity that is produced by this vaccine. The present study was undertaken with following specific object: To isolate and identify *p.multocida* from naturally infected chicken and to determine on farm immunogenicity of adjuvant fowl cholera vaccine produced by LPVRPC.

MATERIALS AND METHODS

Experimental design

The present research was conducted during the period of July 2010 to December 2010.Fowl cholera vaccine as prepared by LPVRPC at BAU was investigated immunogenicity for measured in term of production of antibody in vaccinated chicken determined by PHA test. Seven weeks aged Hy-sex chicken (white and brown) were selected for these experiment. These chicken were divided into two groups- (Vaccinated group and Unvaccinated group) under this vaccinated group collection of blood prior to vaccination from Hy-sex brown A and Hy-sex white B. Primary vaccination were done at 6 weeks of age in Hy- sex brown A and Hy-sex white B. Post primary vaccination bleeding occurred at 10 weeks of age (A_1 and B_1) or 30 days after primary vaccination and Prebooster bleeding at 11weeks of age (A_2 and B_2) or 37 days after primary vaccination. Booster vaccination of Hy-sex brown A and Hy-sex white B were done at 12 weeks of age. Post booster bleeding at 16 weeks of age (A_3 and B_3) or 74 days after primary vaccination. Then collection of sera and PHA test was performed.

Passive haemagglutination (PHA) test

The test was used to determine the antibody titers in chickens and was performed according to the methods described by Tripathy*et al.* (1970a), the sensitivity of PHA test depends upon the use of soluble antigens. In this case, capsular antigens (soluble antigen) of *P. Multocida* were coupled to chemically modified erythrocytes (sheep erythrocytes) and then antigen-coated erythrocytes readily react with specific antibodies and results in haemagglutination.

Microtitre plate method

The procedure of the PHA test was followed according to the method described by Tripathy *et al.,* (1970).An amount of 50 µl of PBS was first poured in each well up to 8th well of horizontal row of microtitre plate. 50 µl of test serum was added in the 1st well. Two fold dilutions of serum ranging from 1: 2 to 1: 256 were made by transferring 50 µl of the mixture from the 1st well to 2nd well and thus continuing successively up to the 8th well from where an excess amount of 50 µl of the mixture was poured off. A volume of 50 µl 0.5% somatic antigen sensitized hRBC was taken in each of the eight wells. The Control system, horizontal row of microtitre plate (9th well: equal volume of 50 µl of normal serum and PBS and10th well: equal volume of 50 µl of sensitized tanned RBC and PBS). The content of the wells of the test system and control were mixed by gentle agitation of the microtitre plate and kept at room temperature for 4 to 5 hours.

The PHA titre was the highest dilution of test sera were complete haemagglutination occur due to the reaction of specific antibody and antigen sensitized tanned HRBC. The results were recorded by deposition of a diffuse thin layer of clumping of RBC on the bottom of the wells, which indicated HA positive, and a compact buttoning with clear zone indicated HA negative. The reciprocal of the highest dilution of sensitized tanned HRBC was considered as titre of the serum.

RESULTS

PHA antibody titer

The PHA antibody titres of the serum obtained from the chicken's belonging to group A1, A2, A3, B1, B2 and B3 are presented in Table-1. The pre-vaccination mean PHA titer were <4±0.00 in sera of chickens of all groups. After 4 weeks of primary vaccination the mean PHA titres were 96.00±34.21 in A1 and 96.00±34.21 in B1 group. Prebooster vaccination PHA titres were 88.00±33.12 in A2 and 96.00±34.21in B2 group. After 4 weeks of booster vaccination the mean PHA titres were 104.00±33.12 in A3 and 104.00±33.12 in B3 group. The mean PHA titres in chickens of unvaccinated control group C were <4±0.00.

Table 1. Mean PHA titres of sera of chickens vaccinated and revaccinated with fowl cholera vaccine through SC route as determined by t-test

Groups	Schedule	PHA titer (Mean ±SE)	P value
A1	Primary vaccination	96.00±34.21	
A2	Prebooster vaccination	88.00±33.12	
A3	Postbooster vaccination	104.00±33.12	
B1	Bleeding at post Primary vaccination	96.00±34.21	0.934
B2	Bleeding at Prebooster vaccination	96.00±34.21	NS
B3	Bleeding after booster vaccination	104.00±33.12	

Level of significance: NS (P>0.05); Legends: PHA=Passive hemagglutination
Mean= Geometric mean of 8 birds; SE = Standard error. NS=Not significant

DISCUSSION

Vaccination is one of the most important methods of prevention of Fowl cholera. This study was undertaken proximately with a view to evaluate the immune responses following usual schedule of vaccination at farm. The immunogenicity was studied by the determination of the serum antibody titre by passive haemagglutination (PHA) test suggested by Carter (1955). PHA test was conducted to determine the humoral immune response of the serum of chickens having been inoculated at 7 weeks aged birds as per the method described by Carter (1955), and Chang (1987) but slight modification was done as suggested by Mondal*et al.,* (1988), Sarker*et al.,* (1992), Siddque*et al.,* (1997), Supar *et al.,* (2002), Akand *et al.,* (2004) and Chowdhury (2008).The prevaccination PHA titres of sera samples of all vaccinates and control birds was found with a mean of <4.00±0.00 that was closely related with Mondal*et al.,* (1988). After 4 weeks of primary vaccination the mean PHA titres were 96.00±34.21 in A1 and 96.00±34.21 in B1 group. Prebooster vaccination PHA titres were 88.00±33.12 in A2 and 96.00±34.21in B2 group. After 4 weeks of booster vaccination the mean PHA titres were 104.00±33.12 in A3 and 104.00±33.12 in B3 group. The mean PHA titres in chickens of unvaccinated control group C were <4±0.00.In this present study, it was observed that group A3 and B3 produced comparatively slightly better immune response than group A1, A2, B1 and B2 and group A1 and A3 produced comparatively better immune response than group A2. There were several limitations of this study such as antibody titres could not determine and compared by ELISA.

Due to short study period immune response of vaccine could not studied elucidated through various routes. *P. multocida* used as antigen in case of microplate agglutination test were identified tentatively by cultural, staining and biochemical test. That could have been identified by molecular characterization such as polymerase chain reaction (PCR).

In conclusion, Fowl cholera vaccine produced by Livestock and Poultry Vaccine Research and Production Centre (LPVRPC) at BAU induced a good level of antibody in layer chicken at farm level. The vaccine produced higher level of antibody when booster dose was given after primary vaccination.

REFERENCES

1. Akand MSI, Choudhury KA, Kabir SML, Sarkar SK and Amin KMR, 2004.Development of washed cell fowl cholera vaccine in Bangladesh. International journal of poultry science, 3: 534-537.
2. Carter GR, 1955. Studies on *P.Multocida* Inhibition Antibody haemagglutination test for the identification of serological types. American journal of veterinary research, 16: 481-484.
3. Chang CF, 1987. In vitro assay for humoral and cell mediated immunity to pasteurellosis in ducks. Journal of Chinese Society of Veterinary Science, 13: 67-74.
4. Choudhury KA, Amin MM, Sarker AJ AH MR and Ahmed AR, 1987. Immunization of chickens against fowl cholera with oil-adjuvanted broth culture vaccine. Bangladesh veterinary journal, 21: 63-73.
5. Choudhury KA, Amin MM, Rahman A, and Ali MR, 1985. Investigation of natural outbreak of fowl cholera. Bangladesh Veterinary Journal, 19: 49-56.
6. Chowdhury MMH, 2008. Study on the comparative efficacy of fowl cholera vaccine prepared with the recent isolate of *P. Multocida* and the commercial one, MS thesis of Department of Microbiology and Hygiene, Bangladesh Agricultural University, Mymensingh.
7. Das PM, Rajib DMM, Noor M and Islam MR, 2005. Retrospective analysis on tile proportional incidence of poultry diseases in greater Mymensingh district of Bangladesh. In proceeding of 4[th]Inlernalional Poultry show and seminar from February 28 to March 2, 2003 held in Bangladesh China Friendship Conference Centre. Agargaon. pp, 35-39.
8. Derieux WT, 1978. Response of young chickens and turkeys to virulent and avirulent *Pasteurella multocida* administered by various routes. Avian Disease, 22: 131-139.
9. Michael A, Geier E, Konshtok R, Hertmanl and Markenson J, 1979. Attenuated live fowl cholera vaccine. I. II. Laboratory and field vaccination trials in turkeys and chickens. Avian Disease, 23: 878-885.
10. Mondal SK, Choudhury KA, Amin MM, Rahman MM and Sarker AJ, 1988. Immune response in chickens induced by alum precipitated fowl cholera vaccine.Humoral immune response. Bangladesh Veteterinary Journal, 22: 63-69.
11. OIE, 2004. Manual of standards for diagnostic test and vaccines. Second Edition.
12. Rahman MK, 2003. Determination of efficacy of experimentally developed formalin killed fowl cholera vaccine via different routes of vaccination. MS thesis of Department of Microbiology and Hygiene, Bangladesh Agricultural University, Mymensingh.
13. Sarker AJ, Amin MNH and Hossain WMA,1992.Testing and quality control of poultry vaccines and its monitoring in the field. Bangladesh Agricultural University Research Progress, 6: 249-257.
14. Siddique AB, Rahman MB, Amin MM and Rahman MM, 1997. Antibody titre on chicks following pigeon pox virus inoculation. Bangladesh Veterinary Journal, 14:12-14.
15. Suman MSR, 2002. Study on the comparative efficacy of experimentally prepared formalin and heat killed fowl cholera vaccines. MS thesis submitted to the Department of Microbiology and Hygiene, Bangladesh Agricultural University, Mymensingh.
16. Supar, Setiadi Y, Djaenuri, Kurniasih N, Poerwadikarta B and Sjafei, 2002. The development of fowl cholera vaccine: II. Pathogenicity and vaccine protection of *P. multocida*local isolates in experimental ducks. Journal of Tm Ternakdan Veterinary, 6: 120-125.

17. Tripathy DN, Hanson LE and MyersWL, 1970a. Passive haemagglutination test with fowl pox virus. Avian Disease, 14: 29-38.

18. Tripathy DN, Hanson LE and Myers WL, 1970b. Detection of fowl pox virus antigen in tissue culture by fluorescent antibody technique. Avian Disease, 14: 810-812.

19. Wu JR, Shieh HK, ShienJH, Gong SR and Chang PC, 2007. Protective immunity conferred by recombinant *P. multocida* lipoprotein E (PlpE). Vaccine, 25: 4140-4148.

21

ASSESSMENT OF MICROBIAL LOAD IN MARKETED BROILER MEAT AT MYMENSINGH DISTRICT OF BANGLADESH AND ITS PUBLIC HEALTH IMPLICATIONS

M. Mahmudul Hasan[1], SM Lutful Kabir[1]*, Nazmul Hoda[2] and M. Mansurul Amin[1]

[1]Department of Microbiology and Hygiene, Bangladesh Agricultural University, Mymensingh-2202, Bangladesh; [2]Veterinary Surgeon, Bangladesh National Zoo, Mirpur, Dhaka, Bangladesh

*Corresponding author: SM Lutful Kabir, E-mail: lkabir79@gmail.com

---

ARTICLE INFO

Key words

Bacterial load
Broiler meat
Broiler market
Public health

ABSTRACT

An investigation was conducted to assess the bacteriological quality of 30 samples of fresh broiler thigh meat samples sold in different retail markets in Mymensingh. Total viable count (TVC), total coliform count (TCC), total salmonella count (TSC) and total campylobacter count (TCpC) in meat samples of different broiler markets like K.R. at BAU campus, Boyra and Kewatkhali were determined. Mean of TVC, TCC, TSC and TCpC for the K.R. at BAU campus, Boyra and Kewatkhali markets were 5.69, 6.03, 6.17 $\log_{10}$ CFU/g, 4.52, 4.66, 4.69 $\log_{10}$ CFU/g, 3.35, 3.51, 3.61, $\log_{10}$ CFU/g and 2.31, 2.56, 2.66 $\log_{10}$ CFU/g, respectively. It was observed that the mean values of TVC, TCC, TSC and TCpC in case of Boyra and Kewatkhali market exceeded the ICMSF recommendations which may cause alarm to consumer's health. The variation of TVC in meats of different broiler market was significant ($P<0.05$) at 5% level of probability whereas TCC and TSC obtained from meat samples of different markets were not showed significant ($P>0.05$).The mean values of TCpC in meats of three different market were highly significant with 1% level of probability ($P<0.01$).There was no significant correlation found between TVC and TCC ($P>0.05$), but a significant correlation found between TVC with TSC and TVC with TCpC in meats of three different markets respectively. Presence of *Escherichia coli*, *Campylobacter* and *Salmonella* spp. in meats must receive particular attention, as these organisms are responsible for causing harm to public health. Suggestion has been given to improve the present sanitary condition of meat processing to minimize bacterial load.

INTRUDUCTION

Chicken is one of the most widely used delicious and versatile meats in the world largely because its protein is of excellent quality and contains all the essential amino acids needed by man. At present broiler is the cheapest meat source in Bangladesh and it contributes about 30% to the total animal protein for human consumption (Huque, 1996). Although small scale commercial boiler farms are gradually rising due to their production yield in shortest period of time and low investment. But contamination of poultry meat with food borne pathogens remains an important health hazardous issue, because of the practices of handling and management during in slaughtering, cooking or post cooking storage of the product (Javadi and Safarmashaei, 2011). Meat was the first important food that met up the hunger of ancient people living in cave (Johanson et al., 1983). It plays a very vital role in keeping the human body strong in order to provide energy, health and vigour (Rahman, 2000).

Meat obtained from chicken does not only undergo spoilage, but may also frequently been found implicated in the spread of food-borne illnesses. If hygienic care is not maintained during the various stages of slaughter operations and processing, the potential edible tissues get subjected to contamination from a variety of sources within and outside the animal and also from the environment, equipment and operators. Morethan 30 genera of micro-organisms including seven pathogens (*E. coli* O157:H7, *Campylobacter*, *Salmonella*, *Clostridium perfringens*, *Listeria monocytogenes*, *Staphylococcus aureus*, are known to contaminate poultry products. Since poultry meat itself offers an excellent medium for the multiplication of most bacteria, including those that are not inhibited by low temperatures, storage of processed poultry meat is vital and therefore considered only under circumstances which inhibit the multiplication of the initial load of bacteria (Adu-Gyamfi et al., 2012). In recent years, food borne infections and intoxications have assumed significance as a health hazard. Epidemiological reports suggest that poultry meat is still the primary cause of human food poisoning (Mulder, 1999). Poultry meat is more popular in the consumer market because of advantages such as easy digestibility and acceptance by the majority of people (Yashoda et al. 2001). However, the presence of pathogenic and spoilage microorganisms in poultry meat and its by-products remains a significant concern for suppliers, consumers and public health officials worldwide. Bacterial contamination of these foods depends on the bacterial level of the poultry carcasses used as the raw product, the hygienic practices during manipulation and on the time and temperature of storage (El-Leithy and Rashad, 1989). However, the control and inspection during production, storage and distribution are generally rare. It is important to prevent the hazards and to provide a safe and wholesome product for human consumption (Singh et al., 1984). Therefore, an investigation was conducted to assess the bacteriological quality of fresh broiler thigh meat samples sold in different retail markets in Mymensingh.

MATERIALS AND METHODS

Sources, collection and transportation of samples

This experiment was carried out in the Microbiology Laboratory of Bangladesh Agricultural University, Mymensingh from 1 January to 1 April, 2014. All samples were obtained from local retail markets situated in K.R. market at BAU campus, Boyra and Kewatkhali market. The attendant first immersed the slaughtered broilers in a special tank containing hot water for some time. The immersed birds were defeathered traditionally by hand plucking and subsequently evisceration was done using special tricks or techniques. Then cut the thigh region muscle and put into a sterilized container. Samples were collected aseptically in sterile containers and brought to the laboratory within 30 minutes to determine the TVC and occurrences of different microflora gaining access to meat. During transportation the sterile containers were kept cool in iceboxes containing fragments of ice.

Preparation of sample for bacteriological studies

Each of the raw meat samples was macerated in a mechanical blender using a sterile diluent as per recommendation of International Organisation for Standardisation (ISO, 1995). Ten grams of the thigh meat sample was taken aseptically with a sterile forceps and transferred into sterile containers containing 90 ml of 0.1% peptone water. A homogenized suspension was made in a sterile blender. Thus 1:10 dilution of the samples was obtained. Later on using whirly mixture machine different serial dilutions ranging from 10^{-2} to 10^{-6} were prepared according to the standard method (ISO, 1995).

Enumeration of TVC

For the determination of total bacterial count, 0.1 ml of each ten-fold dilution was transferred and spread on duplicate PCA using a fresh pipette for each dilution. The diluted samples were spread as quickly as possible on the surface of the plate with a sterile glass spreader. One sterile spreader was used for each plate. The plates were then kept in an incubator at 37^0 C for 24-48 hours. Following incubation, plates exhibiting 30-300 colonies were counted. The average number of colonies in a particular dilution was multiplied by the dilution factor to obtain the total viable count. The TVC was calculated according to ISO (1995). The results of the total bacterial count were expressed as the number of organism or colony forming units per gram (CFU/g) of meat sample.

Enumeration of TCC

For the determination of TCC, 0.1 ml of each ten-fold dilution was transferred and spread on Mac Conkey agar using a sterile pipette for each dilution. The diluted samples were spread as quickly as possible on the surface of the plate with a sterile glass spreader. One sterile spreader was used for each plate. The plates were then kept in an incubator at 37^0C for 24-48 hours. Growth of the organism was confirmed by the appearance of turbidity. Results were calculated from MPN tables.

Enumeration of TSC

For the determination of total salmonella count the procedures of sampling, dilution and streaking were similar to those followed in total viable bacterial count. Only in case of salmonella count, xylose lysine deoxycholate agar (XLDA) was used. The calculation for TSC was similar to that of total viable count.

Enumeration of TCpC

For the determination of TCpC, 0.1 ml of each ten-fold dilution was transferred and spread on the selective blood base agar with 5% sheep or cattle blood. The diluted samples were spread as quickly as possible on 0.45 mm filter placed on blood agar base agar no 2 with a sterile glass spreader. The plates were then kept in an incubator at 42^0C for 24-48 hours. Following incubation, plates exhibiting 30-300 colonies were counted. The average number of colonies in a particular dilution was multiplied by the dilution factor to obtain the total viable count. The total viable count was calculated according to ISO (1995). The results of the total bacterial count were expressed as the number of organism or colony forming units per gram (CFU/g) of meat sample. In young culture the organism is comma shaped and S shaped. In old culture organisms cling together. Gram –ve colonies were round, smooth and translucent with a dewdrop appearance.

Cultural and biochemical examination of samples

The cultural examination of chicken thigh meat samples for bacteriological analysis was done according to the standard method (ICMSF, 1985). The examination followed detail study of colony characteristics including the morphological and biochemical properties. In order to find out different types of microorganisms in chicken thigh meat samples, different kinds of bacterial colonies were isolated in pure culture from the plate count agar (PCA), Mac Conkey agar (MCA), blood agar (BA) and xylose lysine deoxycholate agar (XLDA) and subsequently identified according to the methods described by Krieg et al., 1994. The isolated organisms with supporting growth characteristics on various media were subjected to different biochemical tests such as sugar fermentation test, indole production test, catalase test, coagulase test, methyl-red and Voges-Proskauer (VP) test. In all cases standard methods as described by Cowan (1985) were followed for conducting these tests.

Statistical analysis of experimental data

The data on TVC, TCC, TSC and TCpC obtained from the bacteriological examination of meat samples of the poultry carcass collected from different area of Mymensingh district were analysed in completely randomised design (CRD) using computer package subjected to Analysis of Variance using SPSS Software (Version 16, 2007) . The differences between means were evaluated by Duncan's Multiple Range Test (Gomez and Gomez, 1984). Correlation between TVC, TCC, TSC and TCpC were also evaluated.

RESULTS AND DISCUSSION

The mean and standard deviation of the TVC in broiler thigh meats of K.R. market, Boyra and Kewatkhali markets are presented in Tables 1, 2 and 3. The variation of TVC in meats of different broiler market was significant (P<0.05) at 5% level of probability as shown in Table 4. The result of TVC in three different retail markets were differed significantly (P<0.05). The maximum and minimum range of TVC in thigh meat recorded at K.R. market, Boyra and Kewatkhali markets were log 6.2, log 6.69, log 6.6 and log 5.25, log 5.28, log 5.77, respectively (Table 5). However the average value of TVC at three markets are log 5.72, log5.93 and log 6.18 as shown in Table 5. In K.R. market the value of TVC was lower than Boyra market but it is highest in Kewatkhali market shown in Tables 1, 2 and 3. The possible cause of this variation in microbial load might be thought to be due to differences in management and hygienic practices. Observation of the investigation revealed the fact that in case of K.R. market, the slaughter hygiene and process of broiler meat production was relatively more hygienic in respect of sanitation and handling systems. The butchers generally are skilled and the consumers are well conscious about risk factors and hazardous elements associated with meat production and handling. On the contrary in Kewatkhali markets these are not so, rather the butchers are unskilled and illiterate and the consumers mostly are poor and do not hesitate to purchase poor quality meat. The results obtained were in close agreement with the findings of Adu-Gyamfi et al. (2012), Anwar et al. (2004) and Abu-Ruwaida et al. (1994).

The mean and standard deviation of the TCC of broiler meat processed at slaughter yards of K.R, Boyra, and Kewatkhali markets are summarized in Tables 1, 2 and 3. The result evaluated in Table 4 revealed that the mean values of TCC in meats of K.R. market, Boyra and Kewatkhali market were not significant (P>0.05). Nevertheless no significant variation was demonstrated between the interactions of the three markets. The interpretation of TCC in three different retail markets were not differed significantly (Table 4).The maximum and minimum range of TCC in thigh meat recorded at K.R. market, Boyra and Kewatkhali markets were log 5.1, log 4.96, log 5.15 and log 4.25, log 4.28, log 4.28, respectively (Table 5). However the average value of TCC at three markets were log 4.67, log 4.62 and log 4.71 as shown in Table 5. These findings were agreement with the observations of Datta et al. (2012) and Abu-Ruwaida et al. (1994), respectively. In a study Altabari and Al-Dughaym (2002) and Ahmad et al. (2013) identified lower coliform count of log 3.8 and log 2.94 CFU/ gm, respectively.

The mean values of TSC in broiler meat of three different area like K.R. market at BAU campus Market, Boyra Market and Kewatkhali Market are summarized in Tables 1, 2 and 3. The mean values of TSC in broiler meat of three different area like K.R. market at BAU campus Market, Boyra Market and Kewatkhali Market were log 3.35 ± 0.57, log3.51 ± 0.22 and log3.61 ± 0.76 CFU/g respectively (Table 4). The variation of TSC in meats of different market area was not significant (P>0.05) presented in Table 4. The interpretation of TSC in three different retail markets were not differed significantly (P>0.05). The maximum and minimum range of TSC in thigh meat recorded at K.R., Boyra and Kewatkhali markets were log 3.9 , log 3.8 , log 4.3 and log 2.75, log 3.25, log 3.1 respectively (Table 5). However the average value of TSC at three markets were log 3.32, log 3.47 and log 3.7 as shown in Table 5. The TSC value in K.R market was lower than Boyra market but it is highest in Kewatkhali market. This signifies the fact that all these meats are more or less handled in the same manner. The findings are also closely related to the findings of several other researchers (Mead et al.1994 and Boonmar et al. 1998).

The mean values of TCpC in broiler meat of three different markets like K.R. market, Boyra and Kewatkhali Markets are summarized in Tables 1, 2 and 3. The mean values of TCpC in broiler meat of three different markets like K.R, Boyra and Kewatkhali Markets were log 2.31±0.16, log 2.56 ± 0.03 and log 2.66 ± 0.07 CFU/g respectively (Table 4). The result presented in Table 4 revealed that the mean values of TCpC in meats of K.R. market, Boyra and Kewatkhali market were highly significant with 1% level of probability (P<0.01) .Similarly this variation of TCpC is observed in meats of different broiler carcass as significant (P<0.05). The value of Total Campylobacter Count in three different retail markets were differed significantly (P<0.01) .The maximum and minimum range of TSC in thigh meat estimated at K.R. market, Boyra and Kewatkhali markets were log 2.56 , log 2.77 , log 3.2 and log 2, log 2.5, log 2.1 respectively evaluated in Table 5. Whatever the average value of TSC at three markets a log 2.28, log 2.62 and log 2.65 evaluated in Table 5. The TCpC value of K.R. market is lower than Boyra market but it is highest in Kewatkhali market. These findings have proximal relationship with the findings of Isohanni (2013) and Park et al. (1981).On the otherhand Bodhidatta *et al.* (2013) found higher TCpC value from fresh broiler meat and was log 2.5 to log 3.1 .The value of TCpC at K.R market of BAU campus is lowest (log 2.31) and highest in Kewatkhali market (log 2.66), the findings are also very much close to Federighi et al. (1995), Bjorkroth et al.(2000) and Shane *et al.* (2000) respectively. A significant and positive correlation was found in TCpC at different retail markets with different broiler meat. A highest result 5.33 log 10 c.f.u. per carcass, respectively noted by Cason et al. (1997).

The result estimated in Figure 1 showed weakly correlated between the TVC and TCC. In this study, viable counts were did not significantly correlated with total coliform count in three market area. The regression equation and correlation coefficient values were, y = -0.1323x + 6.5798., and R^2=0.0088 as shown in Fig. 1. The result shown in figure 2 revealed that the regression was positively correlated with TVC and TSC in different market, where correlation coefficient was R^2=0.0377 and regression equation was y = 0.2419x + 5.1227, respectively. Hence the increase of TVC will enhance by the increase of TSC abruptly the decrease of TVC will be decrease with the TSC count (Figure 2). The result evaluated in figure 3 showed positively and significantly correlated between the viable count and total Campylobacter count in three market place. The regression equation and correlation coefficient values were y = 0.5741x +4.5265 and R^2 = 0.1468, respectively (Figure 3). It seems that a significant relation were found between TVC and TCpC respectively.

Table 1. Enumeration of microbial load in broiler meat obtained from K.R market of BAU Campus.

Place of collection	Sample no.	Microbial load			
		TVC (CFU/g)	TCC (CFU/g)	TSC (CFU/g)	TCpC (CFU/gm)
	1	5.8	4.63	3.9	2.1
	2	5.57	5.0	3.83	2.0
	3	5.4	4.25	3.26	2.34
	4	5.25	4.55	3.38	2.44
	5	5.3	5.1	3.15	2.2
K.R. Market	6	5.8.	4.35	3.36	2.5
	7	6.0	3.9	3.65	2.13
	8	6.1	4.5	2.75	2.47
	9	5.55	4.56	3.2	2.56
	10	6.2	4.4	3.1	2.33
Mean ± SD		5.69± 0.28	4.52± 0.16	3.35± 0.57	2.31± 0.16

All counts are expressed in logarithms and CFU/g of meat.

Table 2. Enumeration of microbial load in broiler meat obtained from Boyra Market.

Place of collection	Sample no.	Microbial load			
		TVC (CFU/g)	TCC (CFU/g)	TSC (CFU/g)	TCpC (CFU/g)
	1	5.9	4.7	3.56	2.6
	2	5.28	4.68	3.35	2.56
	3	5.39	4.89	3.65	2.36
	4	5.8	4.77	3.70	2.77
	5	6.45	4.95	3.68	2.8
Boyra Market	6	6.29	4.45	3.29	2.45
	7	5.93	4.35	3.45	2.5
	8	6.37	4.65	3.43	2.66
	9	6.28	4.88	3.7	2.4
	10	6.59	4.28	3.25	2.53
Mean ±SD		**6.03± 0.49**	**4.66±0.29**	**3.51±0.22**	**2.56±0.03**

All counts are expressed in logarithms and CFU/g of meat.

Table 3. Enumeration of microbial load in broiler meat obtained from Kewatkhali market.

Place of collection	Sample no.	Microbial load			
		TVC (CFU/g)	TCC (CFU/g)	TSC (CFU/g)	TCpC (CFU/g)
	1	6.32	4.66	4.3	2.66
	2	6.45	5.15	3.8	3.1
	3	6.2	4.78	3.4	2.55
	4	5.8	4.65	3.67	2.8
Kewatkhali	5	6.15	4.8	3.56	3.2
Market	6	5.77	5.0	3.29	2.1
	7	6.6	4.36	4.1	2.44
	8	5.98	4.28	3.69	2.5
	9	6.05	4.5	3.10	2.7
	10	6.46	4.76	3.22	2.56
Mean±SD		**6.17±0.09**	**4.69±0.07**	**3.61±0.76**	**2.66±0.07**

All counts are expressed in logarithms and CFU/g of meat.

Table 4. Determination of mean and standard deviation for statistical analysis of microbiological quality of chicken thighs at different retail markets in Mymensingh.

Retail Market	TVC Mean ± SD	TCC Mean ± SD	TSC Mean ± SD	TCpC Mean ± SD
K.R. Market	5.69 ± 0.28[b]	4.52± 0.16[a]	3.35 ± 0.57[a]	2.31± 0.16[b]
Boyra Market	6.03 ±0.49[ab]	4.66 ± 0.29[a]	3.51 ± 0.22[a]	2.56 ± 0.03[a]
Kewatkhali Market	6.17 ±0.09[a]	4.69 ± 0.07[a]	3.61 ± 0.76[a]	2.66 ± 0.07[a]
LSD	0.35	0.28	0.31	0.22
Level of sig.	*	NS	NS	**

* = Single asterisk (*) means Significant at 5% level of probability; ** = Double asterisk (**) means Significant at 1% level of probability; NS = Not significant
In a column figures with same letter do not differ significantly ($p>0.05$) whereas figures with dissimilar letter differ significantly (as per DMRT)

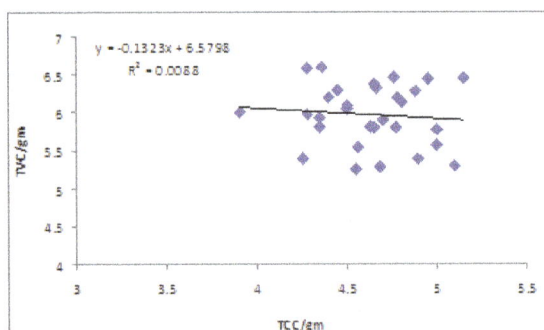

Figure 1. Correlation between TVC and TCC in CFU/g meat of three different retail market

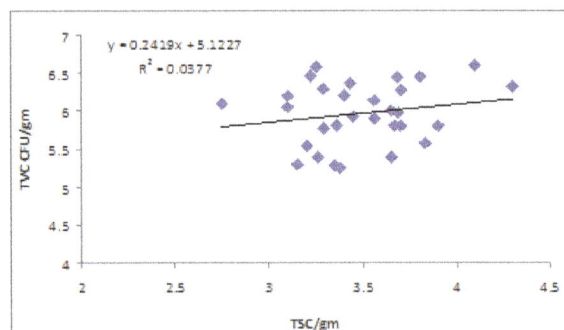

Figure 2. Relationship between TVC and TSC in CFU/g meat of different market area

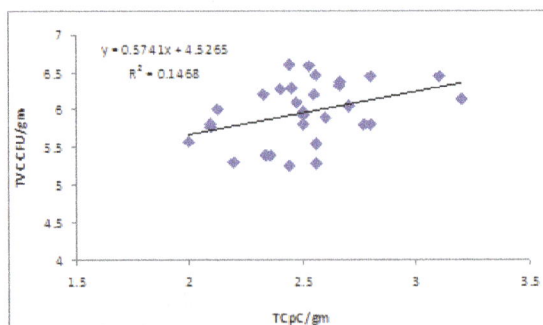

Figure 3. Relationship between TVC and TCpC in CFU/gm meat of different retail market

Table 5. Range of total viable bacteria, coliform, salmonella and campylobacter count in broiler meats obtained from K.R. at BAU, Boyra and Kewatkhali markets.

Source	Examine	TVC			TCC			TSC			TCpC		
		Max	Min	Av.	Max	Min	Av.	Max	Min	Av.	Max	Min	Av.
K.R. market	Thigh	6.2	5.25	5.72	5.1	4.25	4.67	3.9	2.75	3.32	2.56	2.0	2.28
Boyra Market	Thigh	6.59	5.28	5.93	4.95	4.28	4.61	3.70	3.25	3.47	2.77	2.5	2.62
Kewatk-ali market	Thigh	6.6	5.77	6.18	5.15	4.28	4.71	4.3	3.1	3.70	3.2	2.1	2.65

All counts are expressed in logarithms and CFU/gm of meat; Av. = Average

CONCLUSIONS

Bacterial genera identified in this study are known as food borne bacteria which may cause food borne infection and intoxication. The study revealed that the mean values of TVC, TCC, TSC and TCpC of K.R. market was lower than the other two related markets. On the basis of bacterial load K.R. market at BAU campus, Boyra and Kewatkhali markets were graded as A, B and C respectively. Presence of *Escherichia coli*, *Salmonella* and *Campylobacter* in meat must receive particular attention, as these organisms are responsible for causing hazard to public health. These high levels of microbial contamination reflect the poor hygienic quality of poultry meat.

ACKNOWLEDGEMENTS

We are grateful to Professor Dr. M. Mufizur Rahman, Department of Microbiology and Hygiene, Bangladesh Agricultural University, Mymensingh, Bangladesh for critically reading the manuscript.

REFERENCES

1. Abu-Ruwaida AS, WN Sawaya, BH Bashti, M Murad and HA Al-Othman, 1994. Microbiological quality of broilers during processing in a modern commercial slaughterhouse in Kuwait. Journal of Food Protection, 57: 887-892.
2. Adu-Gyamfi A, W Torgby-Tetteh and V Appiah, 2012. Microbiological Quality of Chicken Sold in Accra and Determination of D10-Value of *E. coli*. Food and Nutrition Sciences, 3: 693-698.
3. Ahmad MUD, A Sarwar, MI Najeeb, M Nawaz, AA Anjum, MA Ali and NM Ansur, 2013. Assessment of Microbial load of raw meat at abattoir and retail. The Journal of Animal & Plant Sciences, 23: 745-748.
4. Altabari G and AM Al-Dughaym, 2002. The role of sanitary inspection of meat in relation of food poisoning. In: The Second Annual Scientific Meeting for Environment Hygiene (Meat Hygiene), Riyadh, pp. 180–203.
5. Anower AKMM, MM Rahman, MA Ehsan, MA Islam, MR Islam, GC Shil and MS Rahman, 2004. Bacteriological profile of dressed broilers and its public health implications. Bangladesh Journal of Veterinary Medicine, 2: 69-73.
6. Bjorkroth KJ, R Geisen, U Schillinger, N Weiss, P De-Vos, WH Holzapfel, HJ Korkeala and P Vandamme, 2000. Characterization of *Leuconostoc gasicomitatum* sp. nov. associated with spoiled raw tomato-marinated broiler meat strips packaged under modified atmosphere conditions. Applied and Environmental Microbiology, 66: 3764-3772.
7. Bodhidatta L, P McDaniel and S Sornsakrin, 2010. Case-control study of diarrheal disease etiology in a remote rural area in Western Thailand. The American Journal of Tropical Medicine and Hygiene, 83: 1106-9.

8. Boonmar S, A Bangtrakulnonth, S Pornrunangwong, N Marnrim, K Kaneko and M Ogawa, 1998. Salmonella in broiler chickens in Thailand with special reference to contamination of retail meat with salmonella enteritis. Journal of Veterinary Medical Science, 60: 1233-1236.

9. Cason JA, JS Bailey, NJ Stern, AD Whittemore and NA Cox, 1997. Relationship between aerobic bacteria, salmonellae and campylobacter on broiler carcasses. Poultry Science, 76: 1037-1041.

10. Cowan ST 1985. Cowan and Steel's Manual for Identification of Bacteria (2nd edn.). Cambridge University Press. Cambridge, London.

11. Datta S, A Akter, IG Shah, K Fatema, TH Islam, A Bandyopadhyay, ZUM Khan and D Biswas, 2012. Microbiological quality assessment of raw meat and meat products, and antibiotic susceptibility of isolated *Staphylococcus aureus.* Agriculture, Food and Analytical Bacteriology, 2: 187-194.

12. El-Leithy MA and FM Rashad, 1989. Bacteriological studies on ground meat and its products. Archiv für Lebensmittel Hygiene, 40: 58-61.

13. Federighi M, JM Cappelier, A Rossero, P Cappen and JC Denis, 1995. Assessment of the effect of a decontamination process of broiler carcasses on thermotolerant campylobacters. Sciences-des-Aliments, 15: 393-401.

14. Gomez KA and AA Gomez, 1984. Statistical procedures for agricultural research. John Wiley and sons, Inc. London, UK (2nd edtn)

15. Huque QME, 1996. Improving Skills of the small farmers in poultry management. Poultry Science, 35: 412–418.

16. ICMSF, 1985. Microorganism in foods; samples for Microbiological Analysis: Principles and specific applications. Recommendation of the International Commission on Microbiological Specification for Foods. Association of Microbiological Societies. Toronto, University of Toronto Press.

17. ISO, 1995. Recommendation of the meeting of the subcommittee, International Organization for Standardization, on meat and meat products. ISO/TC-36/Sc-6. The Netherlands. 10-18.

18. Isohanni P, S Huehn, T Aho, T Alter and U Lyhs, 2013. Heat stress adaptation induces crossprotection against lethal acid stress conditions in *Arcobacter butzleri* but not in *Campylobacter jejuni.* Food Microbiology, 34:431-435.

19. Javadi A and S Safarmashaei, 2011. Study of Enterobacteriacea contamination level in premises of poultry slaughterhouse with HACCP system. Journal of Animal and Veterinary Advances, 10: 2163-2166.

20. Johanson L, B Underdal, K Grosland, OP Whelehan and TA Roberts, 1983. A survey of the hygienic quality of beef and pork carcasses in Norway. Acta Veterinaria Scandinavica, 24: 11-13.

21. Mead GC, WR Hudson and MH Hinton, 1994. Use of a marker organism in poultry processing to identify sites of cross contamination and evaluate possible control measures. British Poultry Science, 35:345–354.

22. Mulder RW 1999. Hygiene during transport, slaughter and processing. In: Poultry Meat Science. Poultry Science Symposium Series (Richardson and Mead eds). *CABI Publishing*, 25: 277-285.

23. Krieg NR, JG Holt, PHA Sneath, JT Staley and ST Williams, 1994. Bergey's Manual of Determinative Bacteriology, Williams & Wilkins, Baltimore, Md, USA, 9th edition.

24. Park CE, ZK Stankiewicz, J Lovett and J Hant, 1981. Incidence of campylobacter jejune in frets eviscerated whole market chickens. Canadian Journal of Microbiology, 27: 841-842.

25. Rahman MM, 2000. Fundamentals of Meat Hygiene. Bismillah Farming and Frozen Meat Ltd. Dhaka, pp. 76-101.

26. Shane SM 2000. Campylobacter infection of commercial poultry: Rev Sci Tech Off Int Epiz, 19: 376-395.

27. Singh RD, LN Mandal and JN Pandey, 1984. Isolation of aerobic microorganisms from poultry meat at central poultry farm and retail shop. Patna Poult Guide, 21: 71–74.

28. Yashoda KP, NM Sachindra, PZ Sakhare and DN Rao, 2001. Microbiological quality of broiler chicken carcasses processed hygienically in a small scale poultry processing unit. Journal of Food Quality, 24: 249–259.

Permissions

All chapters in this book were first published in RALF, by AgroAid Foundation; hereby published with permission under the Creative Commons Attribution License or equivalent. Every chapter published in this book has been scrutinized by our experts. Their significance has been extensively debated. The topics covered herein carry significant findings which will fuel the growth of the discipline. They may even be implemented as practical applications or may be referred to as a beginning point for another development.

The contributors of this book come from diverse backgrounds, making this book a truly international effort. This book will bring forth new frontiers with its revolutionizing research information and detailed analysis of the nascent developments around the world.

We would like to thank all the contributing authors for lending their expertise to make the book truly unique. They have played a crucial role in the development of this book. Without their invaluable contributions this book wouldn't have been possible. They have made vital efforts to compile up to date information on the varied aspects of this subject to make this book a valuable addition to the collection of many professionals and students.

This book was conceptualized with the vision of imparting up-to-date information and advanced data in this field. To ensure the same, a matchless editorial board was set up. Every individual on the board went through rigorous rounds of assessment to prove their worth. After which they invested a large part of their time researching and compiling the most relevant data for our readers.

The editorial board has been involved in producing this book since its inception. They have spent rigorous hours researching and exploring the diverse topics which have resulted in the successful publishing of this book. They have passed on their knowledge of decades through this book. To expedite this challenging task, the publisher supported the team at every step. A small team of assistant editors was also appointed to further simplify the editing procedure and attain best results for the readers.

Apart from the editorial board, the designing team has also invested a significant amount of their time in understanding the subject and creating the most relevant covers. They scrutinized every image to scout for the most suitable representation of the subject and create an appropriate cover for the book.

The publishing team has been an ardent support to the editorial, designing and production team. Their endless efforts to recruit the best for this project, has resulted in the accomplishment of this book. They are a veteran in the field of academics and their pool of knowledge is as vast as their experience in printing. Their expertise and guidance has proved useful at every step. Their uncompromising quality standards have made this book an exceptional effort. Their encouragement from time to time has been an inspiration for everyone.

The publisher and the editorial board hope that this book will prove to be a valuable piece of knowledge for researchers, students, practitioners and scholars across the globe.

List of Contributors

Md. Abdullah-Al-Mahmud
Additional Veterinary Surgeon, Upazila Livestock Office, Ullapara, Sirajganj

SM Shariful Hoque Belal
Veterinary Surgeon, District Veterinary Hospital, Sirajganj

Md. Alamgir Hossain
Department of Medicine, Sylhet Agricultural University, Sylhet, Bangladesh

Bayzer Rahman, Tahmina Begum, Yousuf Ali Sarker, Mahmudul Hasan Sikder and Md. Abdul Awal
Department of Pharmacology, Faculty of Veterinary Science, Bangladesh Agricultural University, Mymensingh-2202, Bangladesh

Md. Quamrul Hassan and Sukumar Saha
Department of Microbiology and Hygiene, Faculty of Veterinary Science, Bangladesh Agricultural University, Mymensingh-2202, Bangladesh

Nebash Chandra Pal, Syed Mohammad Bulbul, Zannatul Mawa and Muslah Uddin Ahammad
Department of Poultry Science, Faculty of Animal Husbandry, Bangladesh Agricultural University, Mymensingh-2202, Bangladesh

Md. Nuruzzaman Munsi, Md. Ershaduzzaman, Md. Osman Gani, Md. Moinuddin Khanduker and Md. Shahin Alam
Goat and Sheep Production Research Division, Bangladesh Livestock Research Institute, Savar, Dhaka-1341, Bangladesh

SHM Faruk Siddiki
Faculty of Veterinary Medicine and Animal Science, Bangabandhu Sheikh Mujibur Rahman Agricultural University, Gazipur, Bangladesh

Mohammad Golam Morshed
Upazilla Veterinary Hospital, Chauhali, Sirajganj, Bangladesh

Mst. Sonia Parvin and Lutfun Naher
Department of Medicine, Faculty of Veterinary Science, Bangladesh Agricultural University, Mymensingh, Bangladesh

M. Ariful Islam
Department of Medicine, Faculty of Veterinary Science, Bangladesh Agricultural University, Mymensingh-2202, Bangladesh

ABM Jalal Uddin and Maksudur Rashid
Department of Livestock Services, Bangladesh

Md. Anwarul Khan, Md. Abdul Awal and Md. Abdus Sobhan
Department of Pharmacology

AWM Shamsul Islam
Department of Parasitology, Faculty of Veterinary Science, Bangladesh Agricultural University, Mymensingh-2202, Bangladesh

KBM Saiful Islam
Department of Medicine and Public Health, Sher-e-Bangla Agricultural University, Dhaka, Bangladesh

Md. Ershaduzzaman and Md. Nuruzzaman Munsi
Goat and Sheep Production Research Division, Bangladesh Livestock Research Institute, Savar, Dhaka 1341, Bangladesh

Md. Humayun Kabir and Sompa Das
Conservation and Improvement of Native Sheep through Community Farming and Commercial Farming, Bangladesh Livestock Research Institute, Savar, Dhaka 1341, Bangladesh

Md. Hazzaz Bin Kabir
Department of Microbiology and Parasitology, Sher-e-Bangla Agricultural University, Dhaka, Bangladesh

Mst. Kamrunnaher Akter, ABM Jalal Uddin and Maksudur Rashid
Department of Livestock Services (DLS), Bangladesh

Fahima Binthe Aziz, Md. Bazlar Rashid and Mahmudul Hasan
Department of Physiology and Pharmacology, Hajee Mohammad Danesh Science and Technology University, Dinajpur- 5200, Bangladesh

Tahmina Begum, Mahbub Mostofa and Md. Abdul Awal
Department of Pharmacology, Faculty of Veterinary Science, Bangladesh Agricultural University, Mymensingh-2202, Bangladesh

Bayzer Rahman and Sukumar Saha
Department of Microbiology and Hygiene, Faculty of Veterinary Science, Bangladesh Agricultural University, Mymensingh-2202, Bangladesh

Shah Mohammad Toaha, Bazlur Rahman Mollah and Muslah Uddin Ahammad
Department of Poultry Science, Faculty of Animal Husbandry, Bangladesh Agricultural University, Mymensingh-2202, Bangladesh

Nahid Nawrin Sultana and Mahbub Mostofa
Department of Pharmacology Veterinary Surgeon, Upazilla Livestock Office, Haluaghat, Mymensingh

Soheli Jahan Mou
Department of Surgery and Obstetrics, Faculty of Veterinary Science, Bangladesh Agricultural University (BAU), Mymensingh-2202, Bangladesh Veterinary Surgeon, Upazilla Livestock Office, Islampur, Jamalpur

Md. Abdur Rahman
Concultant Veterinarian, Veterinary Teaching Hospital, BAU, Mymensingh-2202, Bangladesh

Md. Mostafijur Rahman, Khondoker Jahengir Alam and Md. Shah Alam
Department of Pathology and Parasitology, Faculty of Animal Science and Veterinary Medicine, Patuakhali Science and Technology University, Babugonj, Barisal 8210

Md. Mahmudul Hasan and Monalisha Moonmoon
Faculty of Animal Science and Veterinary Medicine, Patuakhali Science and Technology University, Babugonj, Barisal 8210, Bangladesh

Shamir Ahsan, M. Ariful Islam and Md. Taohidul Islam
Department of Medicine, Faculty of Veterinary Science, Bangladesh Agricultural University, Mymensingh-2202, Bangladesh

Farukul Islam, Sifat Hossain Joya and Md. Shamsul Hossain
Bangladesh Agricultural University, Mymensingh-2202, Bangladesh

Muhammad Omar Faruque
Bangladesh Livestock Research Institute, Saver, Dhaka, Bangladesh

Rafiqul Islam and Fowjia Ferdous
Department of Livestock Services, Dhaka, Bangladesh

Md. Mizanoor Rahman
Department of Livestock Services (DLS), Farmgate, Dhaka, Bangladesh

Khan Shahidul Huque, Nani Gopal Das and Md. Yousuf Ali Khan
Bangladesh Livestock Research Institute (BLRI), Savar, Dhaka-1341, Bangladesh

Mohammad Asaduzzaman and Abdul Gaffar Miah
Department of Genetics and Animal Breeding

Ummay Salma, Hussein Suleiman Ali and Md. Abdul Hamid
Department of Animal Science and Nutrition, Faculty of Veterinary and Animal Science, Hajee Mohammad Danesh Science and Technology University, Dinajpur, Bangladesh

Maksuda Begum, Jahura Begum, Md. Shamsul Hossain and Farukul Islam
Bangladesh Agricultural University, Mymensingh-2202, Bangladesh

Md. Kamrul Hasan Majumder
Bangladesh Livestock Research Institute, Savar, Dhaka-1341, Bangladesh

Mohammad Monzurul Hasan
Milkvita, College Road, Islampur, Jamalpur, Bangladesh

Md. Mahmodul Hasan Sohel
Department of Animal Science, Faculty of Agriculture, Erciyes University, Kayseri-38039, Turkey

Md. Ruhul Amin
Department of Animal Breeding and Genetics, Faculty of Animal Husbandry, Bangladesh Agricultural University, Mymensingh-2202, Bangladesh

Md. Abdus Sattar Bag, Md. Mansurul Amin, Md.Bahanur Rahman, Yasir Ahammad Arafat, Md. Salim, Imrul Hasan Rasel and Ummay Habiba Majumder
Department of Microbiology and Hygiene, Faculty of Veterinary Science, Bangladesh Agricultural University, Mymensingh-2202, Bangladesh

M. Mahmudul Hasan, SM Lutful Kabir and M. Mansurul Amin
Department of Microbiology and Hygiene, Bangladesh Agricultural University, Mymensingh-2202, Bangladesh

Nazmul Hoda
Veterinary Surgeon, Bangladesh National Zoo, Mirpur, Dhaka, Bangladesh

Index

www.ingramcontent.com/pod-product-compliance
Lightning Source LLC
Chambersburg PA
CBHW082010190326
41458CB00010B/3144